KV-625-297

THE PHYSICAL BASIS OF THE BRITISH ISLES

PHYSICAL BASIS OF THE BRITISH ISLES

Howard Williams & H. P. Jones

MACMILLAN
London · Melbourne · Toronto
ST MARTIN'S PRESS
New York
1967

MACMILLAN AND COMPANY LIMITED
Little Essex Street London WC 2
also Bombay Calcutta Madras Melbourne
THE MACMILLAN COMPANY OF CANADA LIMITED
70 Bond Street Toronto 2
ST MARTIN'S PRESS INC
175 Fifth Avenue New York NY 10010

HOWARD WILLIAMS, B.A., has considerable experience of grammar school teaching and is an examiner in geography at "O" Level. He is head of geography at Chiswick grammar school

H. P. JONES, B.Sc., is lecturer in charge of geography and geology at Harrow Technical College

PRINTED IN GREAT BRITAIN
BY JARROLD AND SONS LTD NORWICH

PREFACE

THE purpose of this book is to explain the physical setting of regions representative of the different landscapes of the British Isles, through the media of

(*a*) photographs of the varied landscapes,
(*b*) the related Ordnance Survey maps,
(*c*) descriptive text, and
(*d*) supplementary sketches, sections, and line diagrams.

Designed for middle and upper forms, the book is a study of the physical geography of Britain through the use of topographical maps and pictures. The 'facing arrangement' of map and photograph—the latter with superimposed identifications of the chief landscape features—will enable the pupil to recognize more easily the three-dimensional presentation of such features on the Ordnance Survey map.

The approach to physical geography is both practical and visual—and has already been tested in the classroom. The carefully graded exercises will train the pupil to acquire the necessary skills and techniques to interpret satisfactorily both map and picture, as well as an understanding and appreciation of the many types of landscape in the home region. The selected Ordnance Survey maps not only illustrate the regional variety of Britain, but also cover the different types of maps and scales usually set in the G.C.E. 'O'-level and C.S.E. examinations.

The first chapters deal very simply with the formation of landscape; the method of presentation of the different landscapes on Ordnance Survey maps, including considerations of scale; and measurement exercises involving area, distance, gradient, and section construction. Suggested approaches to more advanced exercises appear in later chapters, in which the physical geography of the British Isles is considered by an examination of representative landscapes of selected regions. Whilst there is no substitute for first-hand visual experience of land forms and landscape, we feel that this book is the next-best thing, and hope that it will serve as a useful introduction to field work and a stimulus to further field-study exercises.

H. W.
H. P. J.

ACKNOWLEDGEMENTS

OUR thanks are due to the Director-General of the Ordnance Survey for permission to reproduce Ordnance Survey maps, and to the following for supplying and allowing us to reproduce illustrations: Aerofilms Ltd, for Plates IV, XIII, XIX, XX, XXV, XXVII, XXX, and XXXI; Cheltenham Corporation, for Plate XV; J. Arthur Dixon Ltd, for Plate XI; the Director of the Geological Survey and Museum, for Plates II, III, XII, XIV, XVII, XVIII, XXII, XXVIII, and XXIX; Photography Department, Harrow Technical College, for Plate I; C. V. Wade, for Plate VII; Mr C. A. Inglett, for Plate VI; Mr J. Allan Cash, for Plate XXI; Paul Popper Ltd, for Plate IX; Dr J. K. St. Joseph, University of Cambridge, for Plates V, XXIII, and XXIV; Hugh Sibley, Lensart Ltd, for Plate XXVI; Stewarts & Lloyds Ltd, for Plate XVI; and Mr Anthony Williams, for Plate X. The photograph for Plate VIII was taken by H. P. Jones.

We are greatly indebted to Mr L. E. Carroll of Macmillan & Co. Ltd, for his assistance and advice in the preparation of this book.

H. W.
H. P. J.

CONTENTS

O.S. MAP EXTRACTS

LIST OF FIGURES

LIST OF PLATES

THE GEOLOGICAL TIME SCALE

Era=A major division of geological time
Period=A subdivision of an era
Epoch=A subdivision of a period

Flora=Plant life
Fauna=Animal life

Era	*Period*	*Epoch*	*(Million years)*		*Major events*	*Distinctive life*
			Duration	*Age*		
CAINOZOIC (Recent Life)	Quaternary	Holocene Pleistocene	1		Recent deposition Great Ice Age	Man
	Tertiary	Pliocene Miocene	14 20	15 35	Alpine earth movements Deposition of sands, clays, and gravels in E. and S.E. England	Mammals, birds, and flowering plants
		Oligocene Eocene	15 20	50 70	Great igneous activity in North Britain	
MESOZOIC (Middle Life)	Cretaceous	Chalk Greensands Gault, etc.	50	120	Marine deposits in a sea covering much of England	
	Jurassic	Oolites Lias	30	150	Marine deposits of limestone and clay	Age of reptiles
	Trias	Keuper Bunter	40	190	The New Red Sandstone, formed mostly in inland lakes, and an arid climate	Amphibians and primitive plants
PALAEOZOIC (Ancient Life) UPPER	Permian	—	30	220		
	Carboniferous	Coal Measures Millstone Grit Mountain Limestone	60		Hercynian earth movements	Plants
					Swampy and shallow-water deposits	
				280	A shallow-water, extensive sea	
	Devonian and Old Red Sandstone	—	40	320	Marine sediments in S.W. England Inland deposits in arid climate	Fishes
LOWER	Silurian	—	30	350	Caledonian earth movements	Age of invertebrates
	Ordovician	—	50	400	The great Lower Palaeozoic Geosyncline with marine deposits, especially in Wales and the Lake District	
	Cambrian	—	100	500	Much volcanic activity in Ordovician times	
PRE-CAMBRIAN or ARCHAEAN	—	—	At least 1,000	1,500	Probably several great earth movements, igneous activity, and glaciation periods	Scanty remains of sponges No direct evidence of fossil life
UNRECORDED			At least 2,000			

I HOW LANDSCAPE IS FORMED

I The Making of Landscape

Each of the pictures in this book has been chosen to show some typical aspect of the British landscape. Mountain, valley, plain, or coastal scene—all exist in various forms, sometimes extending broadly over many miles but elsewhere showing great diversity within a quite small area.

The landscape that we see today comprises many man-made features—towns and villages; roads, railways, and canals; and even the very fields and hedgerows. But this book is concerned with the natural landforms, their occurrence and appearance, and mode of formation.

How, in fact, is landscape formed? Briefly, it is the product of the many processes of mountain-building—and destruction—which began in the early phases of earth's history and still go on today. Some of these processes are very gradual—the wearing down of a mountain, or the silting up of a harbour. Others, like landslides and volcanic eruptions, are more immediate in their effects on scenery. The more important of the landscape-making events are summarized in the chart opposite, on which a million years is hardly more than the thickness of a line.

2 The Processes of Landscape Formation

All landforms are composed of rocks, i.e. solid substances of the earth's crust, and all have been under continuous attack by the elements of weather since ancient geological times, when the great rock masses were heaved up to form extensive upland areas. Clearly, landforms that have remained prominent features of landscape are composed of rocks which are resistant to the forces of erosion, while the more subdued landforms are of less resistant and more easily eroded rocks. Some landforms of low relief, e.g. coastal plains, flood plains and deltas, owe their origin to the deposition of material derived from upland areas.

The sketch-map of the British Isles (Fig. 3), showing the chief landform areas of Britain, is a simple expression of the relationship between landforms and rock types. This relationship can be studied in more detail in the geological and relief maps of the British Isles in your atlas.

The processes of creation, decay, and re-creation are illustrated graphically in Fig. 1 and in the chart at the top of p. 3.

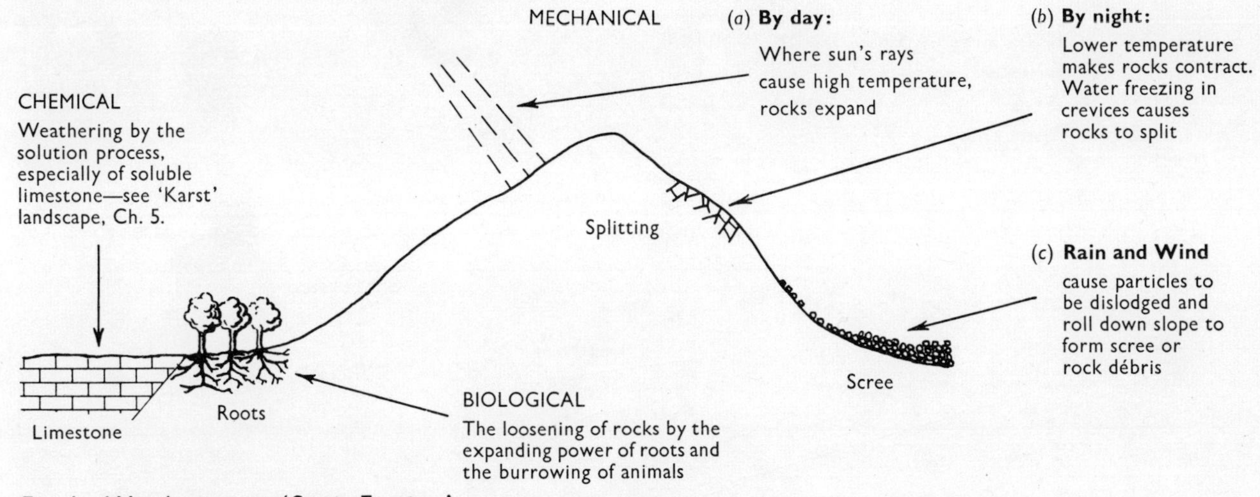

Fig. 1 Weathering, or 'Static Erosion'

This refers to all the processes—e.g. mechanical, chemical, or biological—by which rocks are loosened and decomposed by exposure to the elements

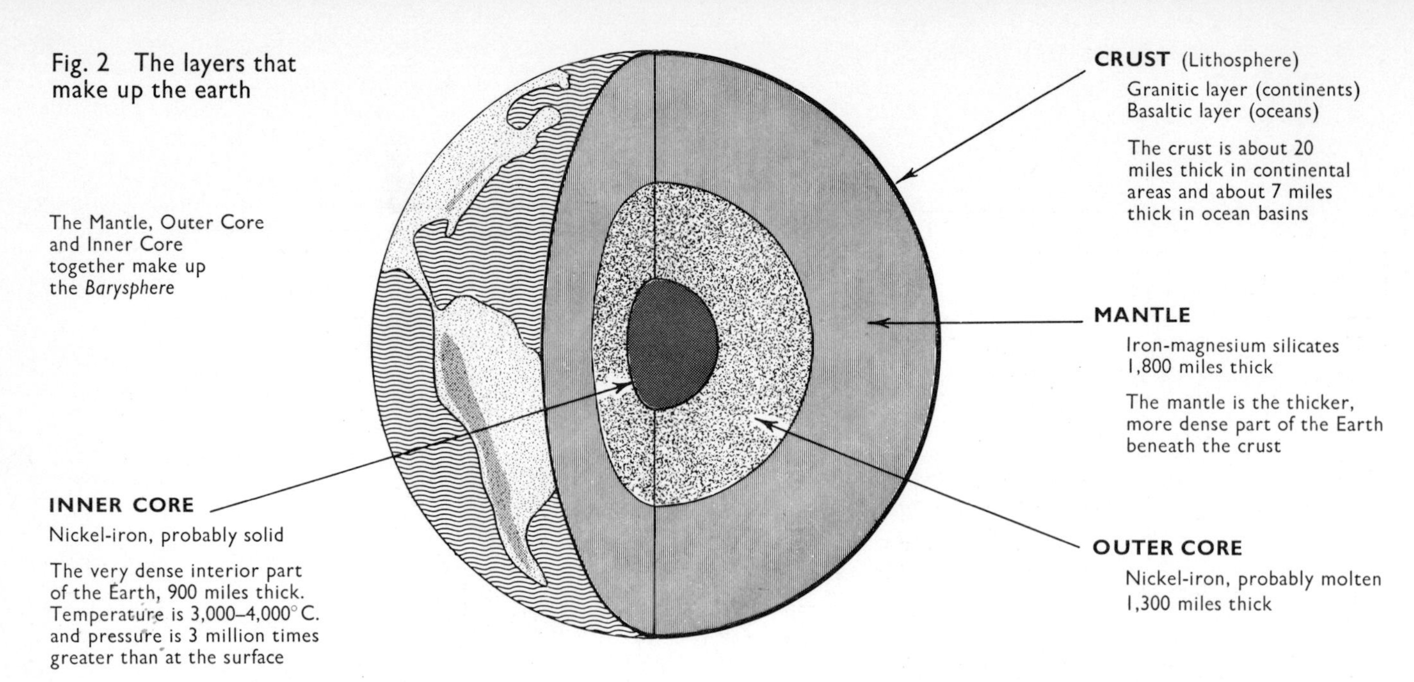

Fig. 2 The layers that make up the earth

Diameter of Earth at Poles = 7,900 miles
Diameter of Earth at Equator = 7,927 miles
Maximum circumference of Earth = 25,000 miles
Land Surface = 29 per cent of total surface area
Water Surface = 71 per cent of total surface area

3 The Origin, Nature, and Classification of Rocks

The earth (Fig. 2) may be considered as consisting of a thin outer skin of rocks (the Lithosphere) surrounding an inner ball of higher density (the Barysphere). The inner ball is regarded as being 'plastic' in nature, in that parts of it are capable of movement when energy-generating rocks, i.e. those containing radioactive elements like uranium and thorium (the source-material of atomic power), liberate a force which puts the more rigid outer skin under tension. Continued tension causes a line of weakness to appear in the crust, through which molten rock or lava (liquefied by intense heat) is transferred, causing the volcanic activity described in Chapter 3.

Rocks may be classified according to age, appearance, and mode of formation into three main groups as shown in the lower table opposite.

EXERCISES

With the aid of your atlas maps showing the relief and geology of the British Isles:

1. Locate and name one area where igneous, sedimentary, and metamorphic rocks occur.
2. (*a*) Name three rock types which are found in that part of south-eastern England lying south of the Thames and east of a line joining London and Southampton.
   (*b*) What prominent relief features do you associate with the rock types named in (*a*) above?
   (*c*) Name three coastal features (e.g. headland) which are the terminal- or end-points on the south-east coast of the features named in your answer to (*b*) above.
3. Attempt a collection of rock specimens of the three main groups shown in the Table opposite—e.g. granite (igneous); sandstone, shale, chalk, and limestone (sedimentary); slate and marble (metamorphic).
4. Complete the following table. The first example has been done for you; do the remainder in the same way.

Area	*Name of prominent relief feature*	*Height of feature*	*Rock type*	*Here is the source of the river*
Dartmoor	High Willhays	2,039 feet	Granite	Dart
North Wales				
Lake District				
Peak District				
The Grampians				

LANDSCAPE

is the product of three processes

1. *Mountain-building*	2. *Denudation*	3. *Deposition*
The process which makes scenery on a large scale	From the Latin word *denudare*, to lay bare; this is the collective term for all the processes and agents responsible for the stripping of the earth's rock cover. Denudation shapes scenery in three ways: (*a*) *Weathering:* the break-up of rocks by action of weather—see Fig. 3 (*b*) *Erosion:* the carving of scenery by rivers, the sea, glaciers, and to some extent by wind (*c*) *Transportation:* the movement of rock débris by the action of the agents listed in (*b*) above	The materials produced by denudation are transported and eventually laid down to form such features as flood plains and deltas

CLASSIFICATION OF ROCKS

1. *Igneous Rocks*	2. *Sedimentary Rocks*	3. *Metamorphic Rocks*
The word 'igneous' is derived from the Latin *ignis* meaning fire, and refers to rocks formed by cooling from a molten state. There are two main types, related to the area or location of cooling: (*a*) *Plutonic Rocks:* these have cooled slowly at depth in the earth's crust. *Example:* granite (*b*) *Volcanic Rocks:* formed by the rapid cooling of lava ejected on to the earth's surface. *Example:* basalt	These are rocks formed by the deposition of land-derived material in seas which once covered the land area in which such rocks occur. *Examples:* sandstones and clays All sedimentary rocks have been derived from igneous rocks which remained above the level of the seas Sedimentary rocks formed (when desert lakes dry up) by the evaporation of water containing substances in solution, such as salt, nitrates, and borax, are known as *Chemical Rocks* Those formed from once living plants, animals, and shellfish, e.g. coal, chalk, peat, and limestone, are referred to as *Organic Rocks*	This term comes from two Greek words, *meta* signifying 'change' and *morphē* meaning 'form'. Such rocks have undergone a change of form as a result of intense heat and/or pressure *Examples:* marble (originally limestone), slate (shale), quartzite (sandstone), and gneiss (from granite)

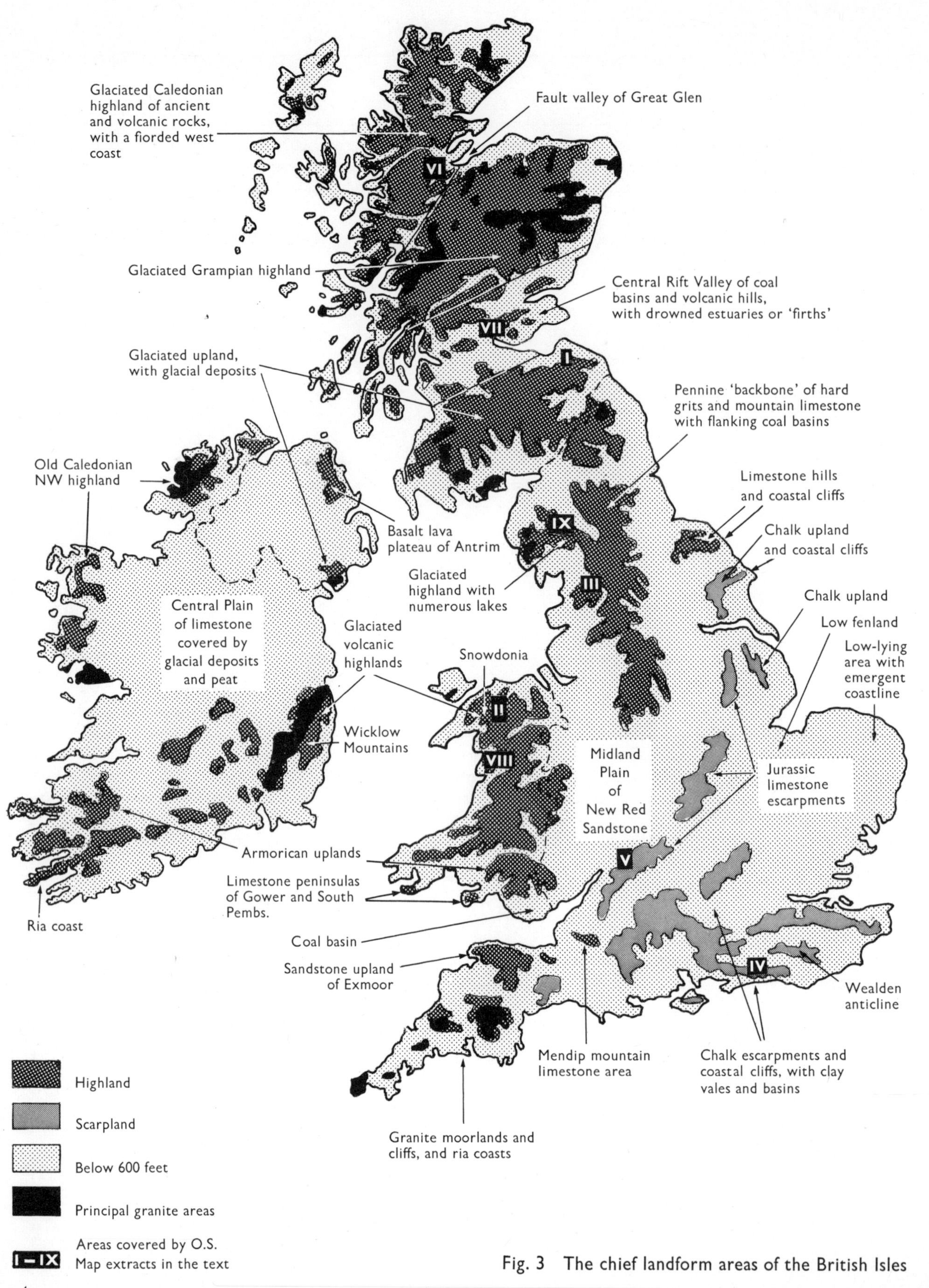

Fig. 3 The chief landform areas of the British Isles

2 LANDSCAPE AND THE ORDNANCE SURVEY MAP

4 Ordnance Survey Maps

Ordnance Survey[1] maps are topographical maps which give a detailed picture of relief instead of the general view of relief shown in atlas maps. Compare the O.S. 1-inch map of part of the South Downs (Map IV, Shoreham, p. 39) with the map of south-eastern England in your atlas. The actual distance on the ground in a straight line between Shoreham and Steyning is 5 miles. On the O.S. map it is 5 inches, whereas in your atlas the distance may be only $\frac{1}{10}$ inch or less. The difference between the O.S. and atlas maps is clearly one of scale.

5 Scale

Scale is the ratio of a distance measured on the map to the actual distance on the ground. Reference to both O.S. and atlas maps will show that scale can be expressed in three ways:

(a) **By a statement,** e.g. 1 inch to 1 statute mile.

(b) **By a representative fraction (R.F.),** which gives scale in numerical form, e.g. 1/63,360, where the numerator 1 (one unit of length on the map) represents 63,360 of the same units of length on the ground.

(c) **By a linear or plain scale.** This is the graphical method of showing scale—a straight line divided into lengths indicating the equivalent distance on the ground.

Note: See also Section 13, Conversion of Scales.

6 Conventional Signs

Landscape features are shown on Ordnance Survey maps by means of the Conventional Signs (p. 9). Note carefully how the symbols used on the O.S. 1-inch map differ from those shown for the O.S. $2\frac{1}{2}$-inch map. The *man-made group of features* includes roads, railways, buildings, and boundaries; the *natural group* consists of relief, vegetation, rivers, and coastal features.

Relief, the most important aspect of our studies, is shown on the O.S. 1-inch map by contours, spot heights, and trigonometrical pillars, i.e. bases for survey instruments. In addition, on the 1-inch Tourist Edition maps relief is also shown by colour layering and hill-shading. On the $2\frac{1}{2}$-inch map, relief is shown by contours, spot heights, and bench-marks. However, one disadvantage of the bench-mark is the need for a separate bench-mark list, as heights are not given alongside the bench-marks on the map.

The contour line, i.e. one joining points of equal height above mean sea-level,[2] is the most effective means of showing relief on O.S. maps. The arrangement of and the spacing between contours varies according to the steepness of slope, the shape, and the height of the relief feature. Study the composite contour map (Fig. 4), and note how the various relief forms are represented by the different contour patterns. The definitions of the relief features should help you to visualize each feature in terms of contour lines.

7 Evidence of Occupations on Ordnance Survey Maps

1. **Farming.** Names of farms are given, and as a general rule the type of farming is related to relief, e.g. sheep farming in upland areas, dairying on water meadows, arable farming in lowlands. However, most farms are of the mixed type.

---

[1] The Ordnance Survey is the Government Department responsible for the survey and publishing of maps of the United Kingdom. The term 'Ordnance' comes from the title of the Duke of Richmond, Master-General of Ordnance (i.e. cannon). In 1791 he made map-making the responsibility of the Army, which discharged this duty until 1895.

The word 'topographical' comes from the two Greek words '*topos*' (place) and '*graphein*' (to describe), but its meaning is restricted today to denote relief only.

[2] Mean sea-level is the average sea-level between high and low tides at Newlyn, in Cornwall, and is referred to as the *Ordnance datum*.

2. **Specialist farming.** Orchards and horticulture are indicated by the appropriate symbols. Often advantage is taken of a southerly aspect and warm soils. Plantations suggest afforestation schemes.
3. **Industrial activities**
   (a) *Mining.* Mines and quarries are shown, e.g. collieries, lead mines, china-clay quarries, clay and gravel pits.
   (b) *Manufacturing.* Factories specifically marked are woollen and paper mills, cement and brick works, hydro-electric plants, and oil refineries (by circular shape).
4. **Tourism.** Suggested by hotels, inns, youth hostels, National Trust areas, mountain-rescue huts, mountain railways, and golf courses.
5. **Other occupations.** Railways, ferries, hospitals, schools, barracks, etc., come under the general heading of 'service industries' and are considered as a single group.

8 Definitions of Landforms

Note: The numbers indicate the positions of the corresponding landforms on the sketch-map (Fig. 4).

Hill Features

1. **Undulating ground:** Gently rising and falling ground.
2. **Knoll:** A low isolated hill (represented by a ringed contour).
3. **Ridge:** A narrow elongated upland with well-marked slopes.
4. **Hog's back:** A symmetrical, steep-sloped hill ridge.
5. **Watershed:** Shown by the dotted line along the summit of the scarp slope. It is the line of high ground separating streams flowing into one river basin from those flowing into an adjoining basin. (A *river basin* is the area drained by a main stream and its tributaries. The collecting area of the river which eventually runs into the main stream is known as the *catchment area.*)
6. **Scarp slope,** and 7. **Dip slope:** The steep and gentle slopes respectively of an escarpment, i.e. a hill ridge with contrasting slopes, which is the main feature west of the river in Fig. 4. This landform is also known as *cuesta*.
8. **Uniform slope:** A regular, even slope, indicated by evenly spaced contours.
9. **Stepped slope:** A short precipitous slope with gentler slopes above and below.
10. **Concave slope:** One which at first is gentle and then becomes steep, in an uphill direction.
11. **Convex slope:** One which at first is steep and then becomes gentle, in an uphill direction.
12. **Spur:** A narrowing extension of upland into low ground; it usually separates one river valley from another.
13. **Interlocking spurs:** Those which alternately overlap, first from one side of a winding stream and then from the other. Such spurs are characteristic of a stream in its source area, i.e. youthful stage.
14. **Shoulder:** A broad level spur with steep slopes above and below.
15. **Saddle:** A depression between two peaks.
16. **Col:** Similar to a saddle, but forms the pass between two summits, often followed by a line of communication, e.g. a road.
17. **Dissected plateau:** A fairly level table-like upland which has been cut into by many streams and rivers.

Glacial Features

18. **Cirque with tarn**
19. **Arête**
20. **Hanging valley with waterfall**
21. **Glaciated valley**

(18–21: for definitions see Chapter 4.)

Valley Features

22. **Dry valley:** A river-formed depression projecting into surrounding upland, but not occupied by a stream.
23. **Gorge:** A long narrow steep-sided opening through upland country.
24. **Flood plain:** The very level plain on both sides of a river, formed of sediment deposited during times of flood.
25. **Bluff:** The steep-sloped edge of a flood plain.

River Features

26. **Confluence:** The meeting of a tributary with the main stream.
27. **Flood-control drainage:** The rectangular pattern of drainage channels designed to prevent flooding.

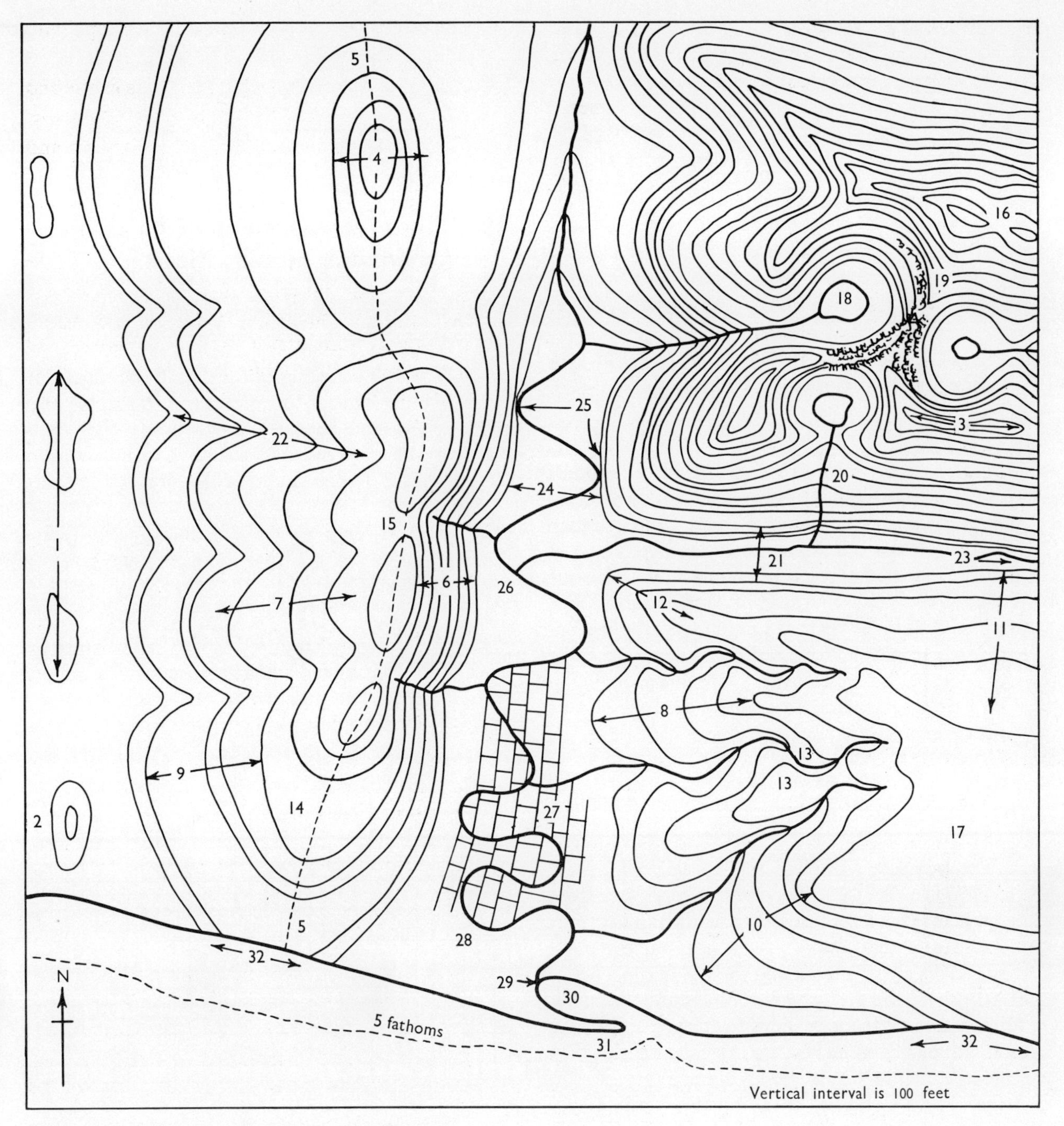

Fig. 4 A composite contour map of landforms

28. **Meanders:** The windings of streams, or the wide sweeping curves of well-developed rivers.
29. **Mouth:** The point where a river reaches the sea.
30. **Estuary:** The drowned mouth of a river, caused by a rise in sea-level, thus creating a broad tidal mouth.
31. **Spit:** A bank or low ridge of sand or gravel projecting into open water, and often found across estuary mouths.

Coastal Features

32. **Cliffs:** Perpendicular rock faces rising from the landward edge of the seashore.

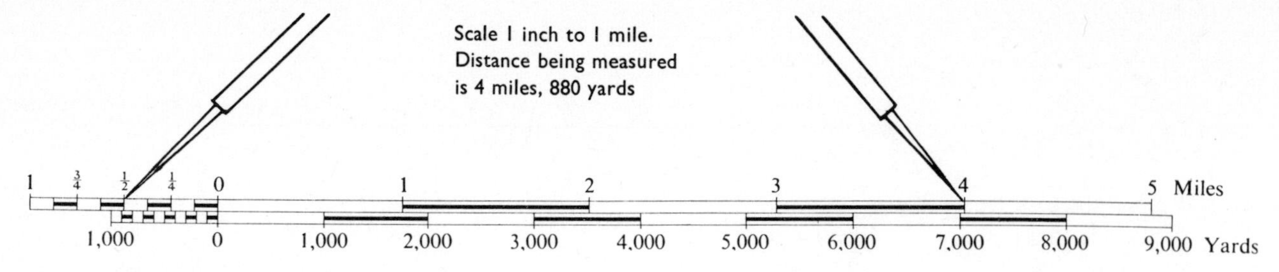

Fig. 5 The measurement of distance on Ordnance Survey maps

9 The Measurement of Distance on Ordnance Survey Maps

(a) *Straight-line Distance*

A quick, accurate method is to use a pair of dividers to 'join' the two points, and then place the open dividers against the scale at the bottom of the map and read off the distance in miles and yards (or as required). Note that the zero point on the map scale is one full division from the left end of the scale. This first full division is further subdivided into units of 220 yards length, and is used to measure the 'left-over' part of the measurement, as shown in Fig. 5. With care, it is possible to estimate accurately to 55 yards or less.

(b) *Curved Distances*

Measurement along curved lines, such as railways, can also be made with dividers—especially for short distances. In this case the dividers are set at a fixed interval of $\frac{1}{2}$ inch—or $\frac{1}{4}$ inch when a railway runs in a series of tight curves. From the starting-point the dividers are pivoted along the railway in a series of half-turns, and the steps must be carefully counted. If, for example, there are nine $\frac{1}{4}$-inch steps, then the total distance is $\frac{9}{4}$ inches, i.e. $2\frac{1}{4}$ miles if the map scale is 1 inch to 1 mile.

For longer distances, break up the winding railway (or road, or river) into a number of short 'straight' sections, which can then be marked off successively along the edge of a sheet of paper. On the edge of the paper mark the starting-point and the end-point of the first of the short sections and then pivot the paper edge to follow the change in direction of the railway, while keeping the end-point in the same position. Mark off the length of the next 'straight' section and continue in the same way until the last end-point is reached. Now set off the total length against the map scale, as with the dividers, and read off the distance.

10 The Measurement of Area on Ordnance Survey Maps

(a) *Measurement of an Irregularly Shaped Area*

(i) Make a tracing of the area to be measured and transfer this outline on to graph paper ($\frac{1}{10}$-inch squares).

(ii) Then, if the map scale is 1 inch to 1 mile, each 1-inch square will represent 1 square mile.

(iii) Add the total of the extra $\frac{1}{10}$-inch squares (100 such $\frac{1}{10}$-inch squares will equal 1 square mile).

Note: The trace line is likely to cut through many of the $\frac{1}{10}$-inch squares. Count as full squares those of which more than half lies *within* the trace line, and ignore those where more than half the square is *outside* the trace line. The method is shown in Fig. 6.

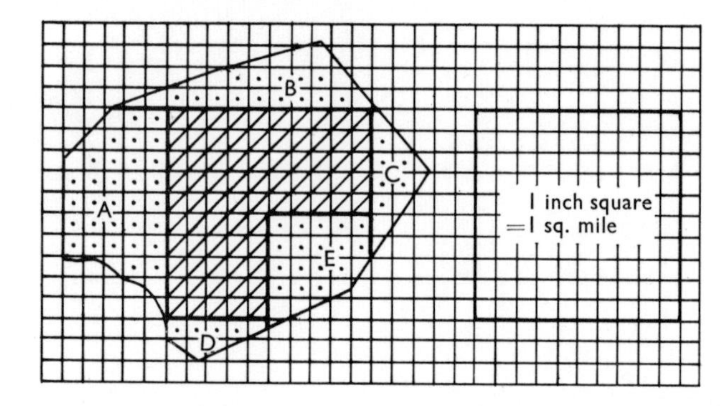

Area of diagram	*No. of whole squares*	*No. of countable part-squares*	*Total No. of countable squares*
Shaded area ($\frac{3}{4}$ sq. in)	75	—	75
Area A	31	5	36
„ B (whole	14	7	21
„ C squares	6	4	10
„ D dotted)	4	3	7
„ E	17	4	21
Area on 1-inch map $=\frac{170}{100}=1\cdot7$ sq. in. ∴ actual area $=1\cdot7$ sq. mls.			170

Fig. 6 The measurement of an irregularly shaped area on an O.S. map

Roads { Ministry of Transport { Motorway — M 1 or A 6 (M)
Trunk { Single & Dual Carriageway — A 33 (T)
Class 1 — A 3055
„ 2 — B 3053
14 ft of Metalling or over (not included above)
Under 14 ft of Metalling { Tarred (not included above) / Untarred „ — TOLL, Gate
Minor Road in towns, Drive or Track (unmetalled) (unfenced roads are shown by pecked lines)
Under Construction
Steep Gradient 1 in 5 or steeper — 1 in 7 to under 1 in 5
Path
Heights in feet above Mean Sea Level { surveyed by levelling .275 / not surveyed by levelling .1091
Triangulation Pillar
Intersection, Lat & Long at 5' intervals (not shown where it confuses important detail)
Public Paths { Footpath (right of way on foot) / Bridleway (right of way on foot and on horseback)
Road used as public path

Railways { Multiple Track / Single „ / Narrow Gauge — Viaduct, Principal Station, Station, Station (closed), Bridge, Level Crossing, Tunnel, Embankment, Cutting, Bridge, Foot Bridge
Mineral Line, Siding or Tramway
Wood
Orchard
Bus or Coach Station
Park or Ornamental Grounds
Youth Hostel
National Trust { always open / opening restricted — NT
Electricity Transmission Line (with pylons spaced conventionally)
Pipe Line (Arrow indicates direction of flow)
Marsh
Glasshouses
Bracken, Heath and Rough Grassland
Quarry
Open Pit
Boundaries { National / County / Civil Parish / Co Boro or County with Civil Parish
Contours at 50 ft intervals — 250
Submarine Contours in Fathoms — 5
Canal, Lake, Weir, Aqueduct, Bridge, Foot Bridge, Lock, Ford, Ferry { Vehicle [V] / Foot [F], Highest point to which Tides flow, Dunes, Cliff, Flat Rock, Beacon, LWM, Sand & Mud, HWM

Church or Chapel { with Tower / with Spire / without Tower or Spire
Wireless or TV Mast
Wind Pump
Windmill (in use) (disused)
Lighthouse — Lightship
Public House PH — Post Office P
Club House CH — Telephone Call Box { PO T / AA A / RAC R
Mile Stone MS
Mile Post MP
Town Hall, Guildhall or equivalent TH
Roman Antiquity (AD43 to AD420) VILLA
Other Antiquities — Castle
Site of Antiquity +
Site of Battle (with date) — 1066

The representation of any other roads, tracks or paths is no evidence of the existence of a right of way

Yds 1,000 — 500 — 0 — 1,000 Yds

REFERENCE

Roads

Ministry of Transport {
Motorway & Class 1 Dual Carriageway — M 4 — A 123
Class 1 — A 123
„ 2 — B 2314 — Fenced — Unfenced
Under Construction

Other Roads — Good, metalled — Poor, or unmetalled
Footpaths — FP Fenced — FP Unfenced
Railways, Multiple Track — Station — Sidings — Road over — Cutting — Tunnel — FB (Footbridge)
„ Single Track — Viaduct — Level Crossing — Embankment — Road under
„ Narrow Gauge
LTE & Glasgow District Subway Stations — Interchange Stations
Aerial Ropeway — Aerial Ropeway
Boundaries { County or County Borough
„ „ County of City (in Scotland)
„ „ „ „ „ with Parish
„ Parish
Pipe Line (Oil, Water) — Pipe Line
Electricity Transmission Lines (Pylons shown at bends and spaced conventionally)
Post Offices (In Villages & Rural Areas only) P — Town Hall TH — Public House PH
Church or Chapel with Tower — Church or Chapel with Spire — Church or Chapel without either
Triangulation Station — on Church with Tower — without Tower
Intersected Point on Chy — on Church with Spire — without Spire — on Building
Guide Post GP. — Mile Post MP. — Mile Stone MS. — Boundary Stone BS — Boundary Post BP
Youth Hostel Y — Telephone Call Box (Public) T — (AA) A — (RAC) R — Antiquity (site of)
Public Buildings — Glasshouses
Quarry & Gravel Pit — Orchard
National Trust Area — Sheen Common NT — Furze
„ „ „ Scotland NTS
Rough Pasture Heath & Moor
Osier Bed — Marsh
Reeds — Well W
Park, Fenced — Spring Spr
Wind Pump Wd Pp.
Wood, Coniferous, Fenced
Wood, Non-Coniferous Unfenced
Contours are at 25 feet vertical interval, shown broken in built up areas. — 100
Brushwood, Fenced & Unfenced
Spot Height 123·

2 Miles — 1 — 0 — 1 — 2 — 3 — Furlongs 4

Ferries — Foot — Vehicle — Sand Hills — LWMMT — Mud — Flat Rock — Slopes — HWMMT — Beacon — Sand — Lightship — Lake — Bridge — Highest point to which Medium Tides flow — Canal — Aqueduct — Lock — Weir — Sand & Shingle — Towing Path — Dam — Ford — FB (Footbridge) — Cliff — Lighthouse

(High & Low Water Mark of Ordinary Spring Tides, in Scotland)

Conventional Signs used on Ordnance Survey maps

Left: Signs for the 1-inch maps (seventh edition)
(Some of these symbols, including, rights of way, are new, and do not yet appear in all maps)

Above: Signs for the 2½-inch maps

(b) *Measurement of Area Consisting of Complete Grid Squares*

Each grid square represents 1 square kilometre, which is 0·386 square mile. So an area covering seven grid squares is 7 square kilometres, or 2·7 square miles.

11 Calculation of Gradient

Gradient is the average slope between two points. Being the ratio between the vertical rise and the horizontal distance between the two points, it is expressed as a ratio, e.g. 1 in 8, that is a rise of 1 foot in height for every advance of 8 feet along the horizontal distance. Gradient is calculated by the simple formula:

$$\frac{\text{Difference in height (in feet) between the two points}}{\text{Distance (in feet) between the two points}}.$$

Example:

Refer to Map V, p. 43 (Cheltenham, Sheet 143).

Q. What is the gradient from the Inn on the road at 919183 to Long Barrow near the summit of the scarp slope in 9317?

A. The height of the road on which the Inn stands is 226 feet. Long Barrow is 850 feet. Thus the difference in height between the two points is 624 feet.

The horizontal distance, in a straight line, between the two points is 1·1 miles=5,808 feet.

$$\text{Gradient}=\frac{624 \text{ feet}}{5{,}808 \text{ feet}}. \quad \frac{624}{5{,}808}=\frac{1}{9{\cdot}3} \doteqdot 1 \text{ in } 9.$$

12 Section Drawing

A section is the outline of the relief of the land between two points on a map. It shows the shape of slopes, valleys, and uplands in 'cut-out' form. The method for constructing a section is as follows:

(*a*) On the Ordnance Survey map join the start- and end-points with a pencilled line; this is the map section line.

(*b*) On graph paper ($\frac{1}{10}$-inch squares) mark the start- and end-points along the top edge.

(*c*) Using the graph paper perpendiculars, transfer these points lower down the graph paper and draw the *horizontal base line* of the section as shown in Fig. 7. The horizontal scale of the section is the same as that of the map.

(*d*) At the left-hand edge of the section base line draw a vertical line, on which the vertical scale will be marked.

Note: The vertical scale must not be the same as the horizontal scale (in this case 1 inch to 5,280 feet), for the highest point on the map is 700 feet and if the vertical and horizontal scale were the same, the resulting section would be of no value. Hence the vertical scale must be exaggerated. Care is necessary here, for if the exaggeration is too great, minor slopes will appear like alpine peaks on the section. In fact no section should be more than $1\frac{1}{2}$ inches in height, or the desired visual impression is lost. A suitable vertical scale for O.S. 1-inch maps is 1 inch to 1,000 feet (exaggeration about 5 times), and for $2\frac{1}{2}$-inch maps, 1 inch to 500 feet (exaggeration about 4 times).

(*e*) Insert the vertical scale; each $\frac{1}{10}$-inch interval represents 100 feet.

(*f*) Again place the graph paper edge along the map section line, marking the position of each contour crossing, and note the height. Then transfer these points to the section and insert at the appropriate height.

(*g*) Join the points by a flowing line and shade as shown in Fig. 7, so that the relief outline is prominent. Annotate the main relief and drainage features, insert the horizontal and vertical scales under the section, and title the diagram, e.g. 'Section west to east from . . . to . . .'.

If a *sketch section* is required, prepare the base and vertical scale lines, examine the line of the section on the map (noting the changes in slope), draw in the relief outline by eye, shade, and title.

Profiles of roads and rivers may be drawn similarly, but as in the case of 'curved distances' (Section 9*b*) the road or river must be broken into a number of straight-line sections. Then, using the edge of a sheet of paper or graph paper, the length of a road or river and the positions of contour crossings are marked. The final result is a 'straightened' road, which becomes the horizontal base-line of the profile. The heights of contour crossings are entered at the appropriate levels against the vertical scale. The points are joined, as before, to give a profile of the road or course of a river.

Intervisibility of points on O.S. maps is determined by drawing lines of sight on the completed section, as shown in the example (Fig. 7).

EXTRACT FROM CHELTENHAM MAP 1-INCH SHEET 143

SECTION WEST TO EAST FROM CHURCHDOWN HILL TO CRICKLEY HILL

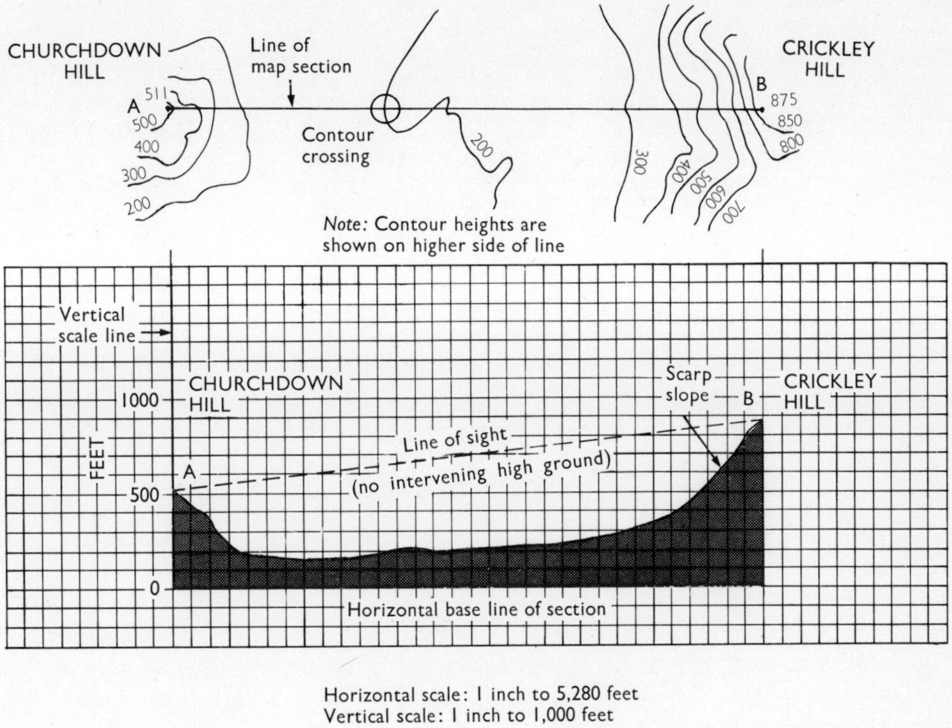

Note: Contour heights are shown on higher side of line

Fig. 7 Section drawing

Horizontal scale: 1 inch to 5,280 feet
Vertical scale: 1 inch to 1,000 feet

13 Conversion of Scales

(*a*) To convert a statement of scale to a representative fraction (R.F.), use the simple formula

$$\text{R.F.} = \frac{\text{M (Map)}}{\text{G (Ground)}}$$

For example, if the scale is 2 inches to 1 mile, what is the R.F.?

$$\text{R.F.} = \frac{\text{M}}{\text{G}} = \frac{\text{2 inches}}{\text{1 mile}} = \frac{\text{2 inches}}{\text{63,360 inches}} = \frac{1}{31{,}680}$$

(*b*) To convert a R.F. to a statement of scale, use the R.F. $\frac{1}{63{,}360}$ as the basis of calculation.

For example,

R.F. $\frac{1}{63{,}360}$ = scale of 1 inch to 1 mile

∴ R.F. $\frac{1}{633{,}600}$ = scale of 1 inch to 10 miles

and R.F. $\frac{1}{6{,}336}$ = scale of 1 inch to $\frac{1}{10}$ mile or 10 inches to 1 mile.

It will be seen that

(i) if the denominator of the given R.F. is *greater* than 63,360 we divide it *by* 63,360 and obtain an answer in miles to the inch.

(ii) if it is *smaller* than 63,360 we divide it *into* 63,360 and obtain an answer in inches to the mile.

Example (i)

The R.F. is $\frac{1}{100{,}000}$. What is the scale in miles to the inch?

1 inch on the map represents 100,000 inches on the ground

$= \frac{100{,}000}{63{,}360}$ miles on the ground = 1·57 miles.

Hence the scale is 1·57 miles to the inch.

Example (ii)

The R.F. is $\frac{1}{25{,}000}$. What is the scale in inches to the mile?

If 25,000 inches on the ground are represented by 1 inch on the map then 63,360 inches on the ground are represented by

$\frac{1 \text{ 'times' } 63{,}360}{25{,}000}$ inches on the map

= 2·534 inches on the map.

Hence the scale is 2·534 inches to the mile.

EXERCISES

Calculate the following:

1. If the scale is 3 inches to 1 mile, what is the R.F.?
2. If the scale is 1 inch to 3 miles, what is the R.F.?
3. If the R.F. is $\frac{1}{250,000}$, what is the scale in miles to the inch?
4. If the scale is 6 inches to 1 mile, what is the R.F.?
5. If the scale is 1 inch to 6 miles, what is the R.F.?
6. If the R.F. is $\frac{1}{600,000}$, what is the scale in miles to the inch?
7. On a French map the R.F. is $\frac{1}{100,000}$. What is the scale in (i) the metric and (ii) the English systems of measurement?
8. On a Japanese map the scale is 1 sŭn to 1 ri. What is the scale in miles to the inch? (*Note:* 1 ri is equal to 129,600 sŭn.)

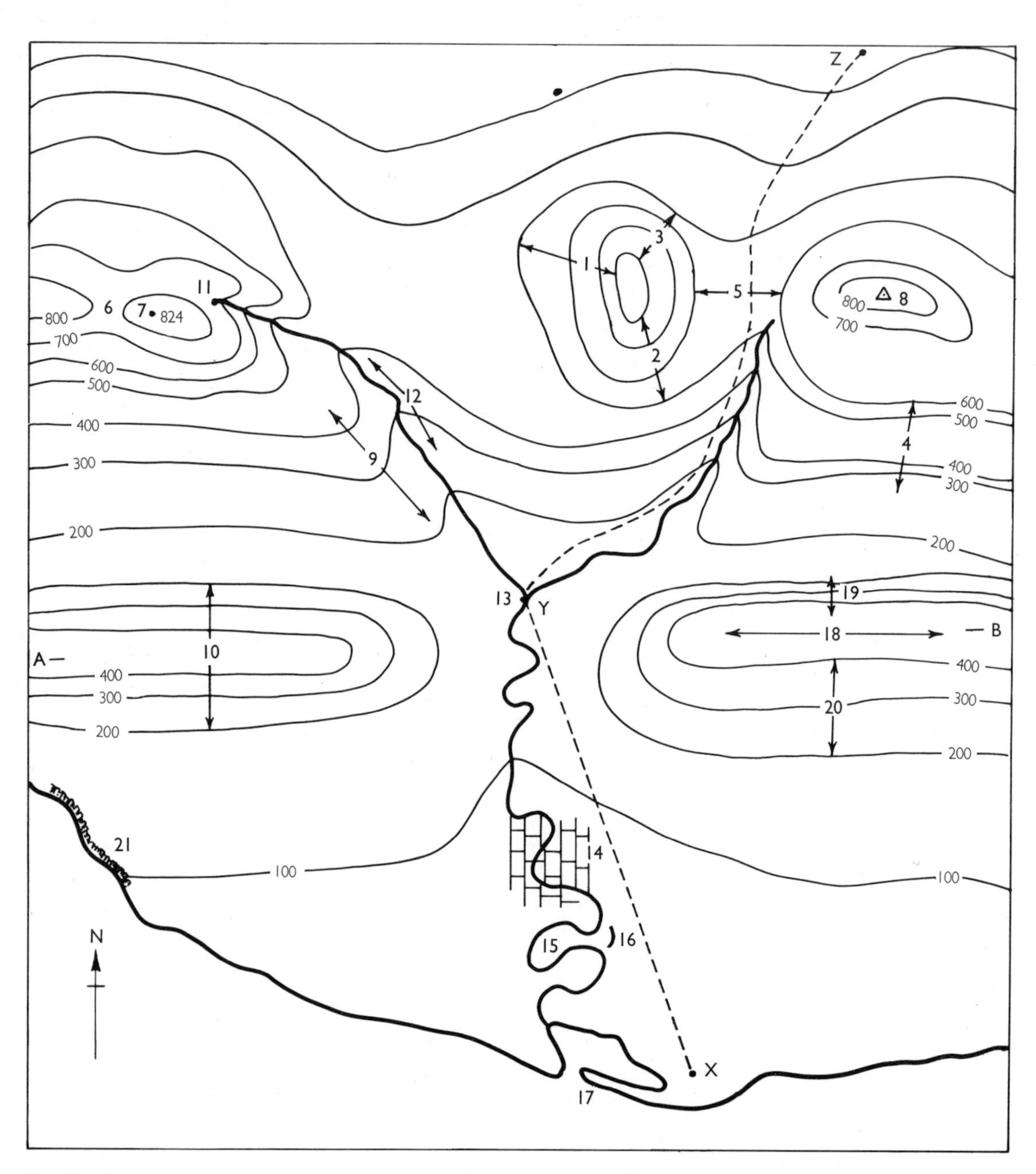

Fig. 8 A contour map exercise

9. The numbers 1–21 in Fig. 8 represent the following features: even slope; concave slope; dip slope; convex slope; scarp slope; col; saddle; river valley; source; drained plain; confluence; hog's back; spur; escarpment; meander; oxbow lake; cliffs; spit; stepped slope; spot height; triangulation pillar.

Draw columns in your exercise book as follows:

Number of Contour feature	*Name of Contour feature*
1	concave slope
2	
3	

Complete the remainder of the columns as shown above from feature 2 to 21.

10. Draw a section (vertical interval $\frac{1}{10}$ inch to 100 feet),
    (i) through the saddle
    (ii) from *A* to *B*, labelling on your section 'water gap' and the other two contour features.

11. (i) From the river confluence point to the spot height 824, find the gradient (assume height of confluence point is 164 feet and scale of map is 1 inch to 1 mile).
    (ii) decide whether the two places are intervisible.

12. Find the area of the hog's back, above the 200-feet contour line.

13. The dotted line is a road joining points *X*, *Y*, and *Z*.
    (i) If the scale is 1/63360, what is the distance from *X* to *Y*?
    (ii) If the scale is $2\frac{1}{2}$ inches to 1 mile, what is the distance from *Y* to *Z*?

3 MOUNTAIN LANDSCAPES

14 How Mountains are Formed

Mountains are extensive areas of high relief, having been uplifted by earth movements during the successive periods shown on the Geological Time Scale (p. viii). Present-day earthquakes and volcanic action, widespread in the world, are evidence of powerful interior forces.

During the periods of gigantic earth movements known as the Caledonian, Hercynian, and Alpine, parts of the earth's crust were subjected to great pressure or tension, thus creating the two processes whereby mountains are formed, namely:

(a) **Folding:** the bending or arching of rock strata of the earth's crust.

(b) **Faulting:** the displacement of strata along a line of weakness in the earth's crust.

On the basis of the mode of formation and of the processes involved, three main types of mountains are recognized: *fold*, *block*, and *volcanic* mountains.

15 Fold Mountains

These are formed by the bending of rock strata of the earth's crust by compression, as shown in Fig. 9*a* and *b*.

Examples of folding in the British Isles are to be seen in the Cemaes Head coastal area in north-west Pembrokeshire, in the Gower Peninsula near Swansea, in the Bude area of Cornwall, and in the extensive anticline of the Pennines. Folding in these areas occurred during the Hercynian period, and such folds are examples of what is known as *old folding*, as opposed to the formation of *young fold* mountains, such as the Alps, formed during the 'Alpine' period. The old fold mountains of Highland Britain (the region of Britain lying north and west of a line joining the mouths of the rivers Tees and Exe) have been subjected to the processes of denudation over a very much longer period than young fold mountains, and are thus less spectacular features of landscape than the Swiss Alps.

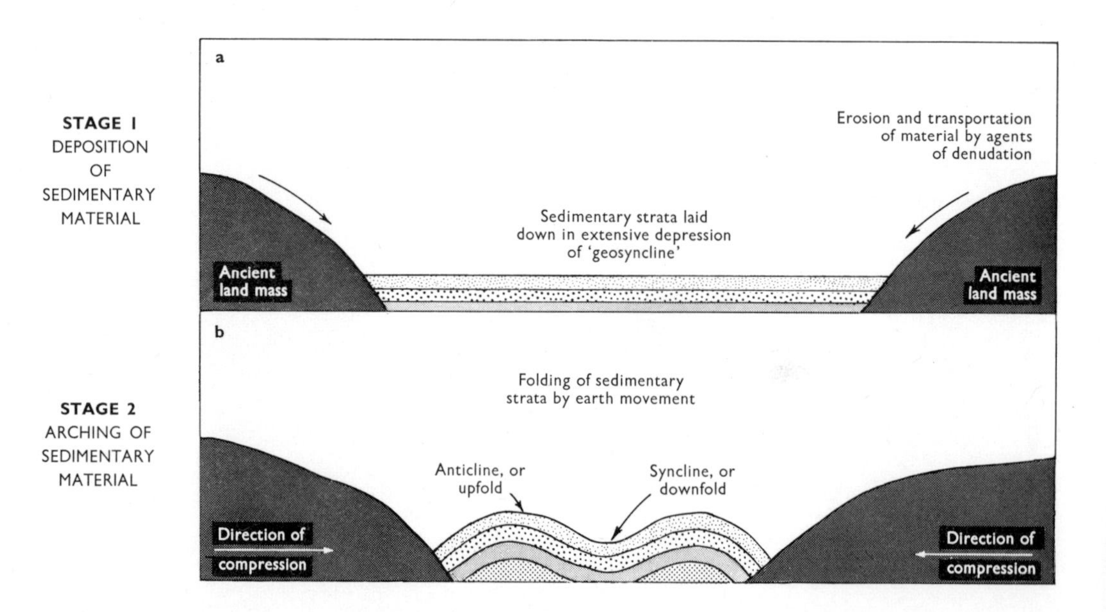

Fig. 9 The formation of fold mountains

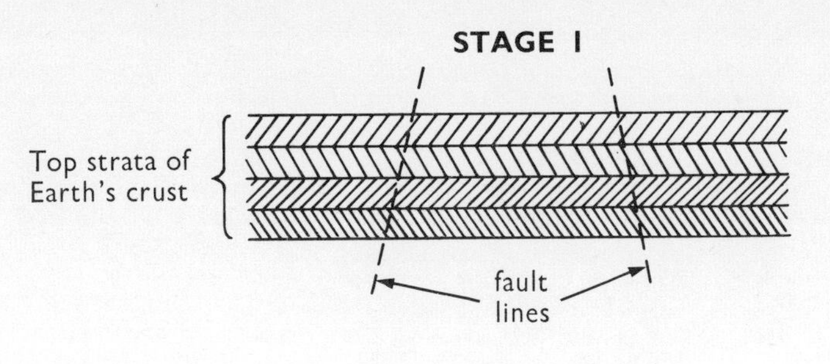

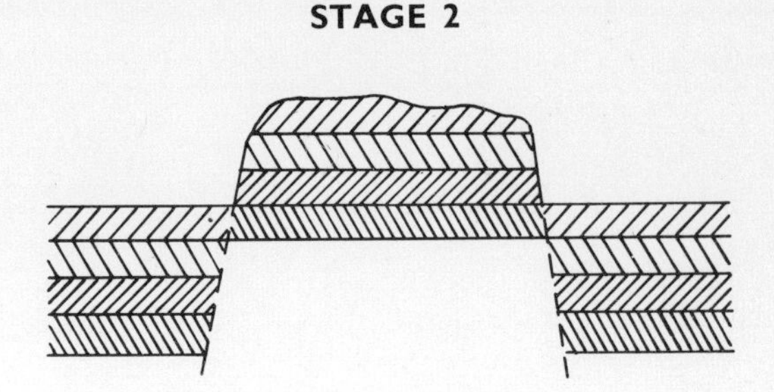

Fig. 10 How a block mountain is formed

Earth's crust under tension; appearance of fault lines

Subsidence of Earth's crust on either side of fault lines

16 Block Mountains

Earth movements causing stresses in the earth's crust produce fracture lines (also known as faults or lines of weakness). A continuation of the earth movement may cause great masses of the crust to subside below the general level or to be raised above it. In this way block mountains are formed. Fig. 10 shows that the block mountain is bounded by faults, and such an example is known as a *horst*. The block mountains of central Tasmania are of this type. There are no examples of horsts in the British Isles, but the North Pennines are a tilted block and the eastern edge of the Eden Valley is formed by the Pennine Fault, which culminates in Cross Fell at an elevation of 2,930 feet, as shown in Fig. 11.

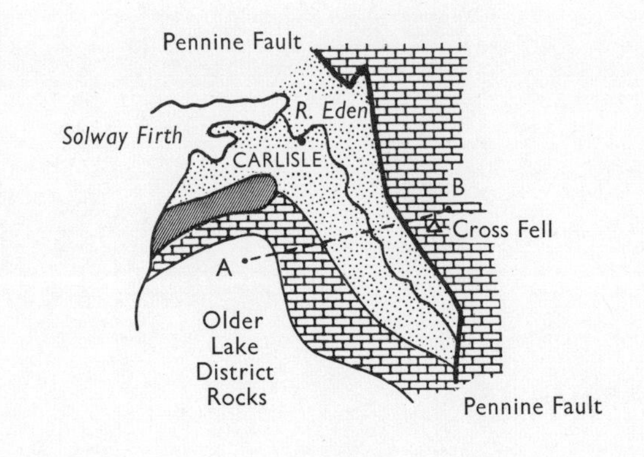

Section A–B

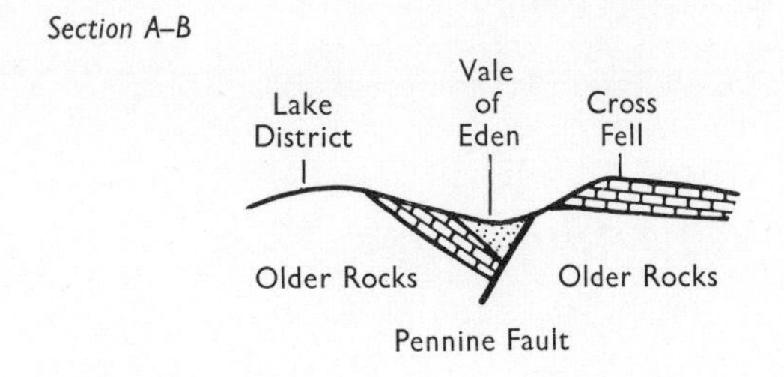

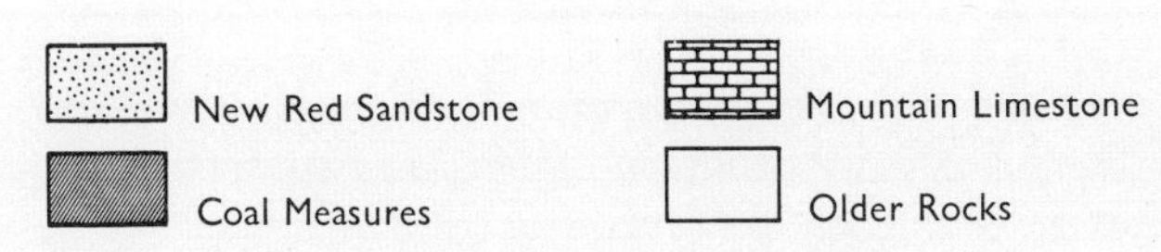

Fig. 11 The north-west Pennines, showing the Pennine Fault

17 Rift Valleys

Associated with block mountain formation is that of rift valleys—elongated tracts of country let down between parallel faults (Fig. 12).

There are no extensive rift valleys in the British Isles like those in the Rhine Valley and East Africa, but an example on a smaller scale is the Central Lowland of Scotland, between the fault along the southern edge of the Grampians from Stonehaven to Helensburgh (the 'Highland Boundary Fault') and that along the northern edge of the Southern Uplands from Dunbar to Girvan (the 'Southern Upland Fault').

The longest fault in Britain—Glen More in the Highlands of Scotland—is not a rift valley, but one formed by a glacier moving along a fault line. The Vale of Clwyd in North Wales, from Ruthin to St Asaph, is a miniature rift valley.

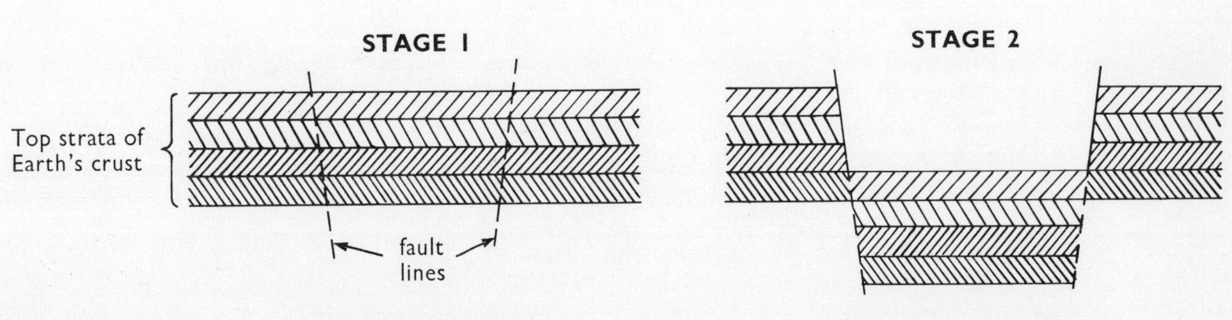

Fig. 12 How a rift valley is formed

Earth's crust under tension; appearance of fault lines

Subsidence of part of Earth's crust between fault lines

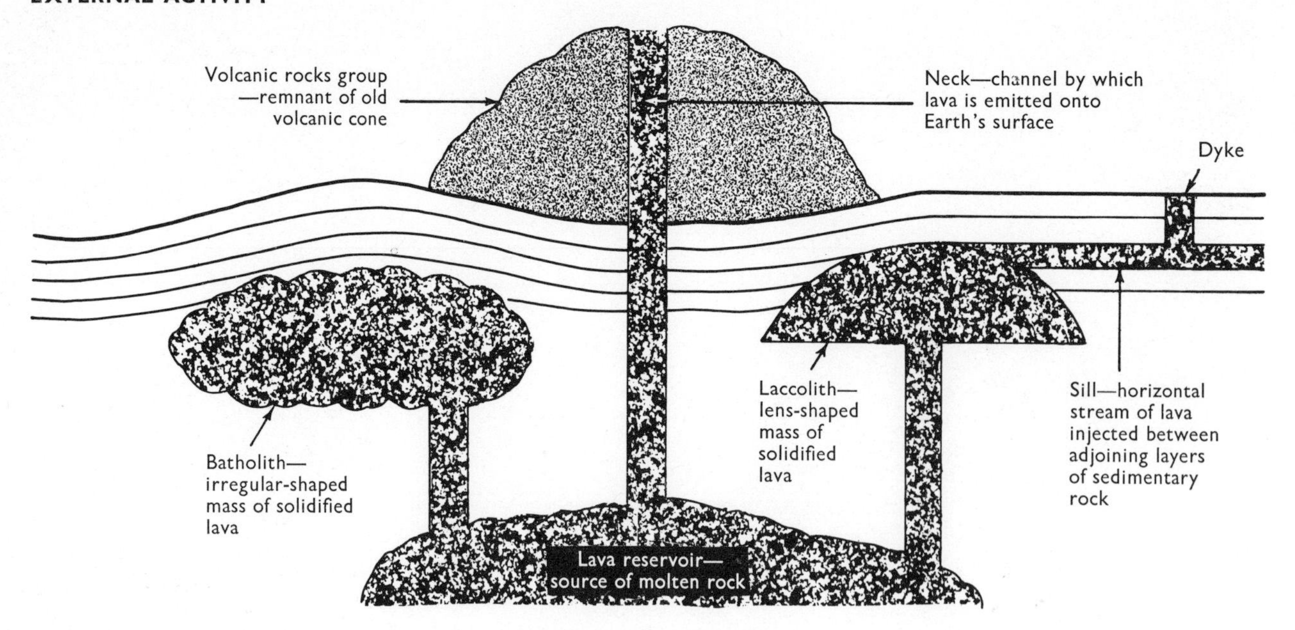

INTERNAL ACTIVITY

Such features become part of landscape only after removal of overlying sedimentary layers by erosion, e.g. Traprain Law, East Lothian (Laccolith); Cader Idris cliffs, North Wales (Sill), overlooking Barmouth estuary; Dartmoor (Batholiths)

Fig. 13 Some of the features associated with volcanic activity

18 Volcanic Mountains

In areas of folding and faulting following disturbances in the earth's crust there may also be a release of molten rock or lava from deep sources within the earth. This transfer of lava *towards* or *onto* the earth's surface via a line of weakness in the earth's crust is known as volcanic action. There are two main types of such volcanic activity:

(a) **Internal,** where lava is thrust into strata beneath the actual surface without breaching the skin of the earth. Such penetration is termed an igneous intrusion. Fig. 13 shows the features associated with internal activity, and these become part of landscape only *after* the overlying strata have been stripped by denudation.

(b) **External,** which occurs when lava is ejected on to the surface. The nature of the activity is determined by the composition of the lava. If it is composed mainly of silica, with little mineral content, it is described as 'acid'. This lava is viscous, not free-flowing, and when ejected during an eruption it tends to pile up, forming convex-shaped cones. But lava with a high mineral and low silica content, i.e. 'basic lava', is more fluid and may flow considerable distances before solidifying.

19 Volcanic Landscape in the British Isles

There are no young fold mountains in the British Isles, and consequently no recent zones of weakness where volcanic activity could occur.[1] What evidence there is of volcanic activity must be related to ancient mountain-building, such as the Caledonian and Hercynian, and the regions affected by these. Here the old volcanic cones or vents have long been masked by deposition or removed by erosion. However, a study of the geological map in your atlas will reveal that volcanic rocks occur in many areas in Highland Britain. Some examples are:

(a) **North Wales.** The 1-inch O.S. map extract of the Snowdon district (Map II, p. 27) shows three mountain groups: Snowdon, the Glyder (6458), shown on Plate I, and Carnedd (6663). All are formed of old volcanic material. Cader Idris, overlooking the Barmouth Estuary, is also a volcanic upland.

[1] This explains, too, the rarity of earthquakes—earth tremors or vibrations in strata at the earth's surface—often induced by volcanic activity when great masses of lava are moved. The most recent earth tremor in England occurred in the Chichester area in September 1963, and was caused by a very slight movement along a fault line in the English Channel.

PLATE I Glyder Fawr volcanic rocks south of Llyn Idwal, North Wales. Llyn Idwal is in the foreground

PLATE II The Whin Sill and Hadrian's Wall

PLATE III Haytor, near Ilsington, South Devon

(b) Lake District. Here the volcanic rock region extends south of the line joining Ullswater and Derwentwater to Ambleside, and includes the mountain areas of Helvellyn, Scaw Fell, Great Gable, and the Langdale Pikes.

(c) Northumberland. North of the River Tyne is an extensive lava sill of dolerite—a coarse basalt—originally injected into the Carboniferous rocks which make up much of the Pennines, and solidifying there. The stripping of the covering rock layers revealed the more resistant Whin Sill—a very prominent landscape feature (Plate II).

(d) South-west Peninsula. Between Totnes and Dittisham the River Dart flows through volcanic lavas. The tors—Hensbarrow Down (St. Austell), Haytor on Dartmoor (Plate III), and Brown Willy on Bodmin Moor—are remnants of batholiths, once covered but later exposed by the very gradual removal of the original rock cover.

(e) East Lothian, Scotland. Traprain Law, four miles east of Haddington, is an example of a 'revealed' laccolith—a lens-shaped mass of solidified lava (Plate IV).

EXERCISES

Map I, East Linton (extract from O.S. $2\frac{1}{2}$-inch Sheet NT 57) and Plate IV, Traprain Law.

1. Draw a contour plan of Traprain Law (scale 5 inches to 1 mile and contour interval of 50 feet) and insert:
   (i) the line of trees and two roads which appear near the bottom edge of the photograph;
   (ii) an arrow to indicate the north direction and one to show the direction in which the camera was pointing.
2. Trace the outline of Traprain Law (take the 400-foot contour to be the boundary) and transfer this on to graph paper. Calculate the area of this relief feature (again note the scale of the map). Measure the length and breadth of Traprain Law and state its height.
3. Using the information already gained, write a paragraph on Traprain Law.
4. From the map, measure the length of the line of trees which runs south-east from the quarry in 5874. What is the length of this line of trees on the photograph? Use this information to calculate the approximate scale of this part of the photograph.
5. Two settlements on the photograph have been marked *A* and *B*. Name them.
6. What evidence is there of the past and present use of Traprain Law?
7. Examine the course of the River Tyne on the map and then answer the following questions:
   (i) In which direction is the river flowing? (The preliminary note on p. 49 will help you to answer this question.)
   (ii) Give two reasons to support your answer to part (i).
   (iii) What are the dots which appear in the river's course in square 5575?
   (iv) Locate and give the map reference of the three weirs on the river's course and explain their purpose.
8. How far is it from the railway station in East Linton to the road junction in Traprain (5975),
   (i) in a straight line, (ii) by road?
9. What is the conventional sign between 585775 and 575776, and why is it necessary?
10. There are many wells and springs in the map area. Suggest two reasons for this.

PLATE IV Traprain Law, East Lothian

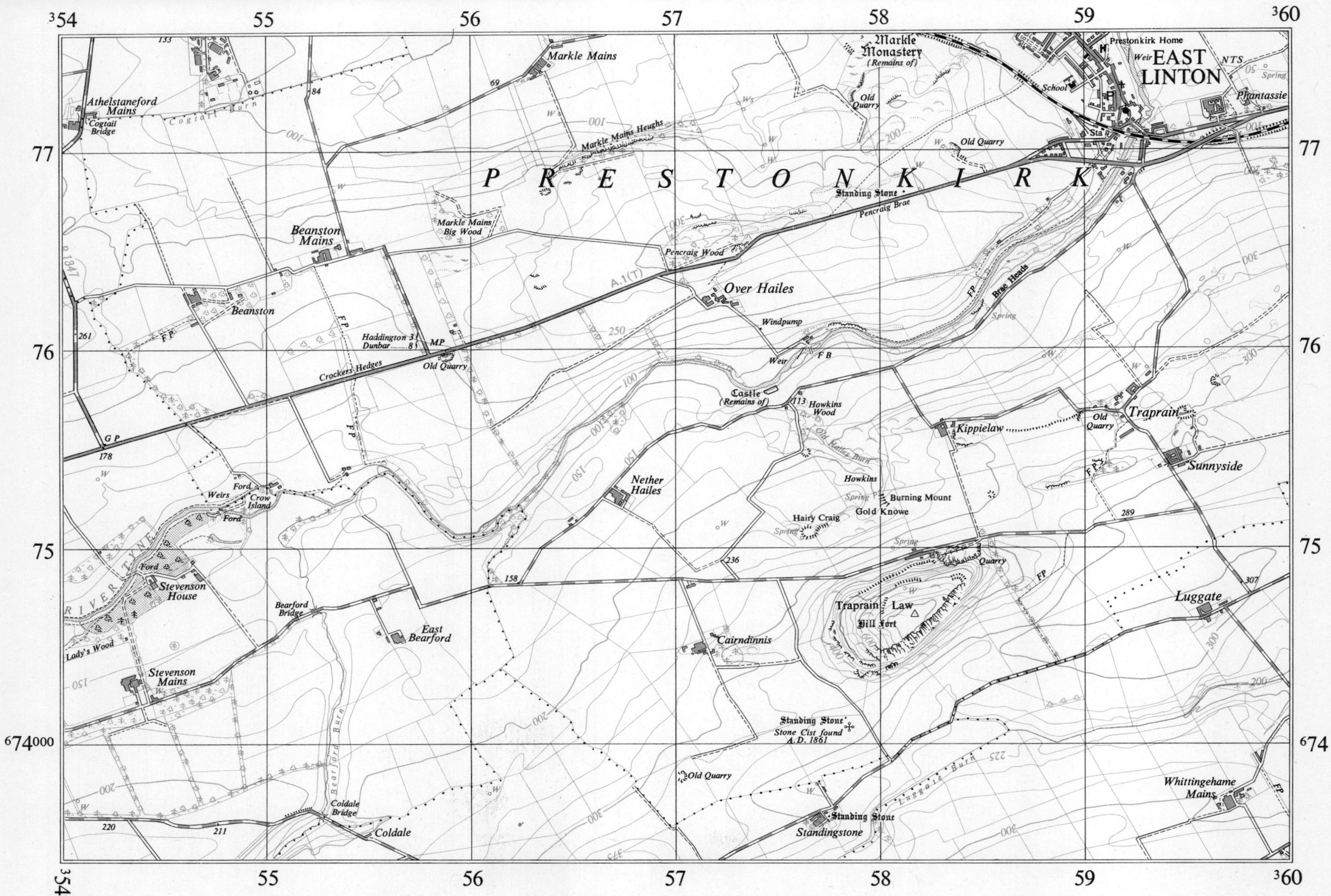

MAP I — East Linton

[Pri]nted by Lowe & Brydone (Printers) Ltd., London

4 GLACIATED MOUNTAIN LANDSCAPES

20 Definitions of Glacial Terms

Ice Age: A period of climatic 'deep freeze', such as the Quaternary Period in the Geological Time Scale (p. viii).

Valley glacier: A very slow-moving 'river' of ice, formed at the valley head by the accumulation of snow—originally in hollows above the snow-line (Fig. 14). Successive layers of snow form first the semi-solid ice, termed *névé*, and then *glacier ice*. This eventually moves down the valley-slope under its own weight and then, armed with rock fragments embedded into its base, becomes a powerful agent of erosion by its abrasive, grinding, and plucking action.

Glaciated valley: A glacier-eroded valley (U-shaped in the section originally occupied by the ice) whose upper slopes form prominent shoulders. The valley has a flat floor bounded by steep walls, and across it there meanders a misfit stream—i.e. one occupying a valley far too large to have been formed by the stream itself.

Ice sheet: An extensive layer of thick ice covering a vast tract of land. (During the Ice Age it covered the North Sea as well.)

Hanging valley: A tributary valley left high above the floor of the main glaciated valley, and formed by the over-deepening of the latter valley by the greater eroding power of the glacier.

Beheaded spurs: Those whose lower slopes have been eroded by the processes named above, thus straightening the valley walls.

Cirques, cwms or corries: Armchair-shaped depressions formed at the head of glaciated valleys by the eroding action of moving ice (Figs. 14 and 18).

Arête: A razor-like divide, separating back-to-back cirques, and formed by the headward erosion of the back walls of the cirques.

Pyramidal peak: A horn-shaped peak formed when an arête is eroded by a battery of cirques isolating a section of the arête, which is then eroded on all flanks.

Moraines: Frost-shattered rock débris and material

STAGE 1 FORMATION OF GLACIER

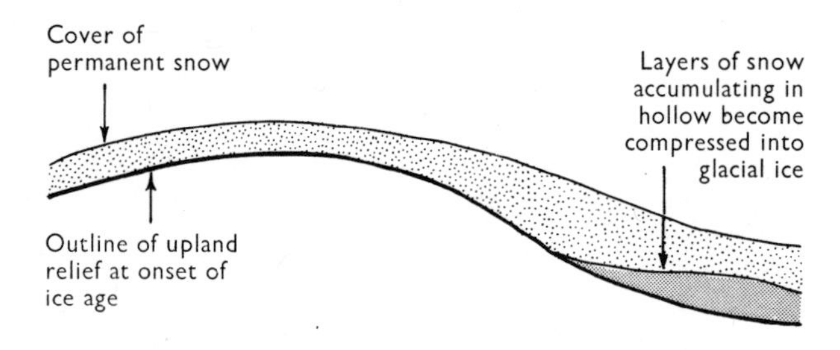

STAGE 2 MOVEMENT OF GLACIER

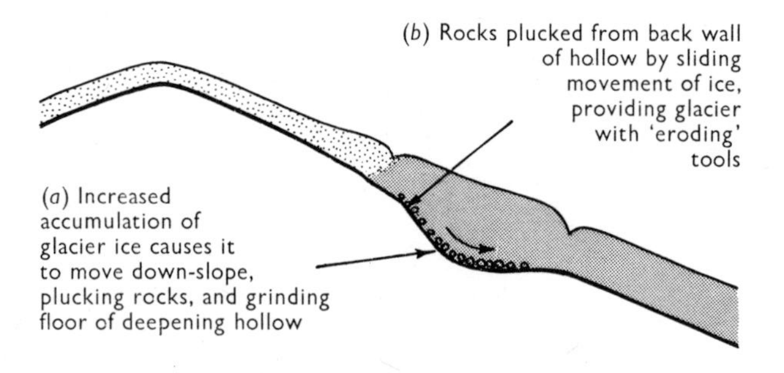

STAGE 3 THE GLACIER-FORMED CWM

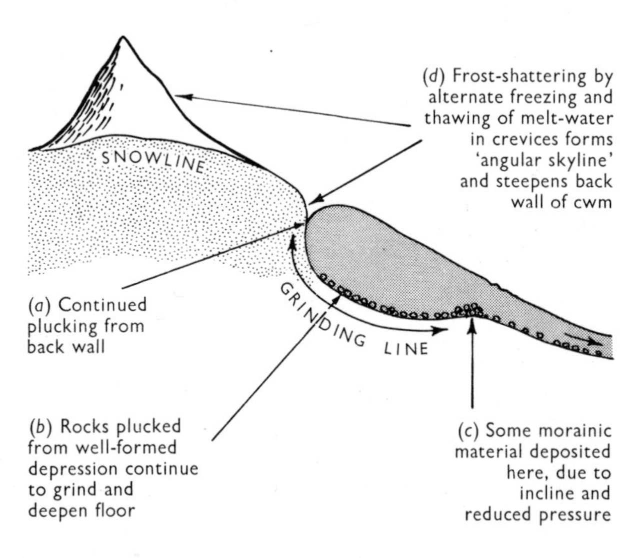

Fig. 14 Stages in the formation of a cwm (cirque or corrie)

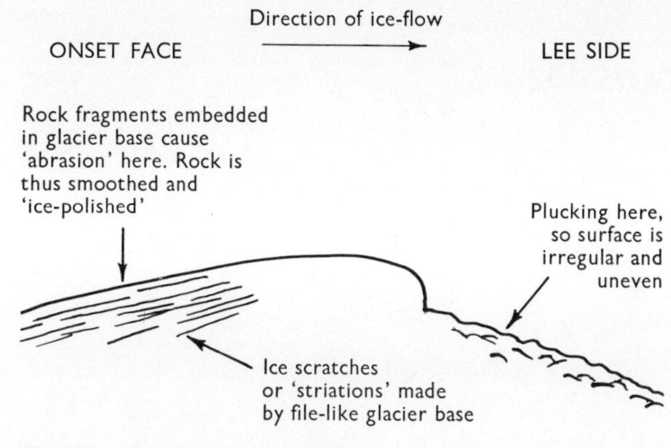

Fig. 15 A roche moutonnée

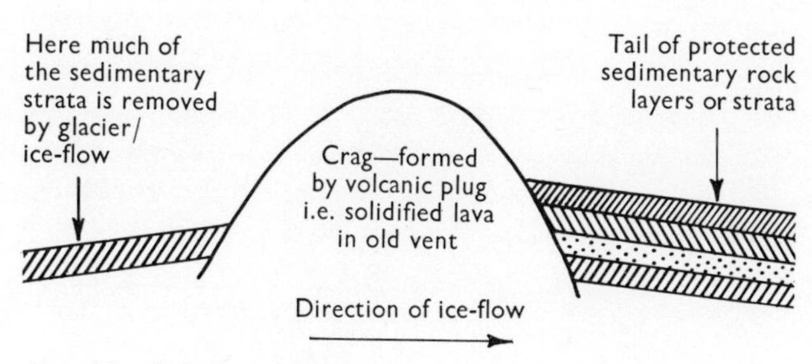

Fig. 16 A crag and tail

derived from the valley sides and floor, transported and deposited by glaciers. Note the several types: *lateral*, *medial*, *ground*, and *terminal*. Extensive spreads of ground moraine are referred to as *boulder clay*.

Drumlins: Long hillocks of boulder clay producing an effect referred to as a 'basket of eggs' type of scenery (Plate V).

Erratics: Rocks which, transported and deposited by glaciers, are 'alien' to the region in which they occur, e.g. the Ailsa Crag rocks from western Scotland found in Snowdonia.

Roches moutonnées: Residual hummocks embedded in the floor of the glaciated valley, smoothed on the face which is 'onset' to the moving glacier and rough and irregular on the opposite side. The name is derived from their resemblance to the rippled sheepskin wigs worn in eighteenth-century France (Fig. 15).

Terminal moraine: Glacial-deposited material marking the point of maximum advance of the glacier.

Outwash plain: That formed by the deposition of fine sands and gravels by melt-water streams in the area beyond the terminal moraine.

Eskers: Long winding ridges of sands and gravels deposited by melt-water beneath the glacier.

Glacial lakes: Lakes which owe their formation to glacial action. There are four types: the *tarn* on the floor of the cirque, the *ribbon lake* dammed by the terminal moraine, the lake occupying a *glacier-scooped basin*, and a type not found in Britain, namely, the *glacier-dammed lake* such as the Marjelen See, dammed by the Aletsch Glacier, in the Bernese Oberland in Central Switzerland (see Chapter 9).

Crag and tail: These occur where a rock outcrop (crag) lying in the path of a glacier is resistant enough to protect softer and less resistant rocks on the lee side. *Example:* the crag on which Edinburgh is built and the tail of sedimentary strata on which Princes Street extends westwards from the Castle (Fig. 16).

Striations: Scratches on ice-scarred rock surfaces, e.g. on the onset face of roches moutonnées (Fig. 15).

Crevasses: Cracks in the surface of the glacier caused by tension when the glacier moves over a well-marked break of slope, or rounds a bend.

Séracs: Pinnacles or pillars of ice formed at intersections of lateral and longitudinal crevasses (Fig. 18).

Icefall: A combination of crevasses and séracs. Note that icefall features occur only where there are existing glaciers, and not in the British Isles (Fig. 18).

21 Glaciated Mountain Landscape in North Wales

During the Ice Age, valley glaciers and ice sheets covered the whole of the British Isles north of a line joining Bristol and London, and considerably altered

PLATE V Drumlins, Nether Lodge, Ribblehead, Yorks.

the then existing landscape by glacial erosion. The diagrams in Fig. 17 show the changes in scenery produced by glacial erosion. Note especially the features named on the diagrams, as well as their definitions in the list of terms.

The photograph of the Nant Ffrancon Valley in Snowdonia (Plate VII) is typical of glaciated uplands in other such regions of the British Isles—the Lake District and the Highlands of Scotland. A study of Map II (O.S. Sheet 107, the Snowdon district) facing the photograph makes it clear that the Nant Ffrancon Valley is the most prominent strip of level relief on the map. The close packing of the contours on either side of the valley floor indicates that this is a glaciated valley, but the walls are not precipitous, as in the Austerdalsbrae Valley (Plate VI), a tributary of the Sogne Fjord in Norway. Can you suggest a reason for this difference in slope?

There are many other features of glaciated landscapes to be observed on the map. These include:

(a) **Glacial lakes,** of the finger or ribbon type, on many of the valley floors, formed as a result of a terminal moraine blocking melt-water streams from a receding glacier, or by the melt-water filling glacier-scooped basins. It is difficult to identify morainic deposits on O.S. Map II, but at 560621 the projecting contour and knoll-like arrangement of contours suggest the presence of moraines which have dammed Llyn (Lake) Padarn.

(b) **Tarns,** occupying the floors of cwms (cirques), indicated by the semi-circular pattern of contours around the tarn.

(c) **Pyramidal peak** of Snowdon hemmed in by a battery of cwms.

(d) **Arête.** There is a splendid example of this type of steep ridge, separating back-to-back cwms, 1 mile north-west of Snowdon summit.

Of interest also is Llyn Ogwen, at the southern end of the Nant Ffrancon Valley. It is thought that in pre-glacial times drainage from the valley now occupied by Llyn Ogwen flowed eastwards into the Afon (River) Llugwy. But glacial over-deepening between Rhaedr Ogwen (648604), where there is a glacier-formed stepped slope, and point 675605 caused rivers such as Nant (Stream) Gwern-y-gar to flow westwards into what is now Llyn Ogwen. This is a good example of a reversal of direction of river-flow by glacial over-deepening.

1 BEFORE THE ADVANCE OF GLACIERS

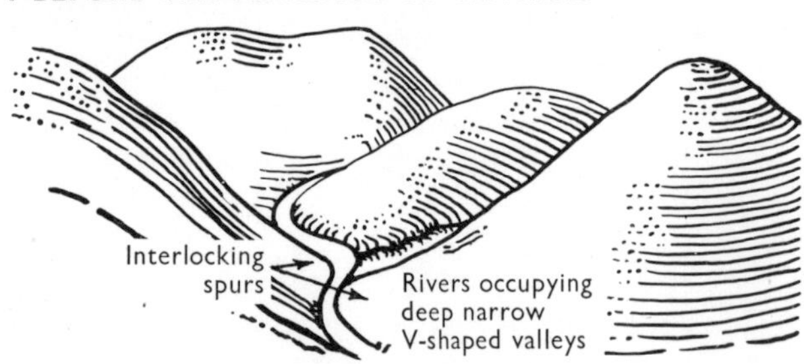

2 DURING GLACIATION

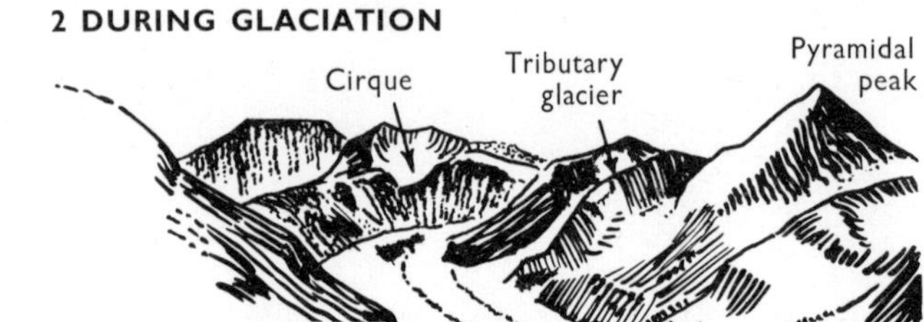

3 AFTER RETREAT OF GLACIERS

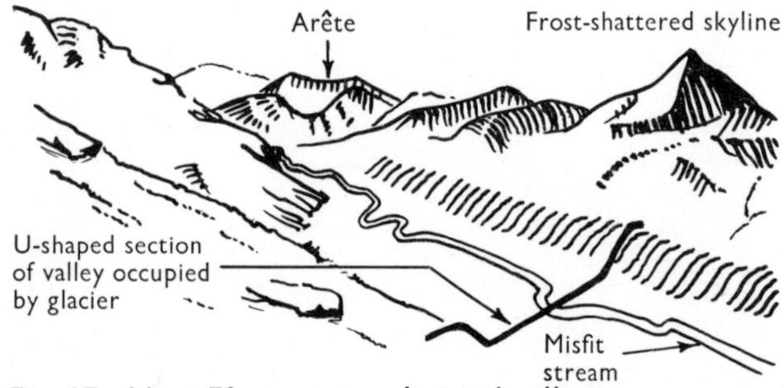

Fig. 17 Nant Ffrancon—a glaciated valley

EXERCISES

Map II (extract from O.S. 1-inch Sheet 107).

1. (*a*) Give the grid reference of:
   (i) the junction of two 'A' roads
   (ii) the junction of an 'A' road and a 'B' road.
   (*b*) Identify the symbols used to show:
   (i) relief in square 6155
   (ii) vegetation in square 5860.
   (*c*) What are the five types of roads shown on the map?
   (*d*) Locate and name three types of railways and briefly state their uses.
2. Examine the following statement: Occupations in the map area include quarrying, tourism, transport, and afforestation. What evidence is there on the map to support this statement?
3. North of the 59 northings grid line, five streams have their sources in tarns. Name two of these tarns.
4. Why is settlement so sparse in this area?

5. (*a*) Give the grid reference of a youth hostel on the map.
   (*b*) What is
      (i) the straight line distance in miles and furlongs between the Llanberis and Bryngwynant hostels?
      (ii) the distance by road?
6. Describe and give reasons for the distribution of woodland in the map area. A suitable approach is:
   (i) state the areas where woodland occurs,
   (ii) give the general upper limit of tree growth,
   (iii) refer to aspects of setting, e.g. high relief, exposure to strong winds, thin soils, and the presence of marshland—all factors which could explain the absence of woodland.

   Note that sandstone areas favour coniferous growth, while chalk country supports lime-loving trees like the beech. The term 'natural vegetation' includes all forms of growth—woodland, heathland, coarse grasses, and rough pasture, as well as marsh vegetation, but not nurseries and orchards, which are man-made.
7. (*a*) Draw a rectangle on a scale of 2 inches to 1 mile to represent that part of the map area between eastings 60 and 67, and north of the 60 grid line and insert:
      (i) the contours for 1,000 feet, 1,500 feet, 2,000 feet and 3,000 feet.
      (ii) the terms peak, ridge, col, and precipitous slope at the appropriate places.
      (iii) the main valley.
   (*b*) What is the map evidence to show that two southward flowing streams are mountain torrents?
   (*c*) What kind of valley is that of Nant Ffrancon?
8. Examine the route of the Snowdon Mountain Railway and
   (i) calculate the average gradient between the station in square 5859 and the Summit Station.
   (ii) State which is the steeper gradient—that between Hebron and Halfway Stations, or that between Clogwin and Summit Stations.
9. (*a*) Explain by a simple sketch the appearance and mode of origin of a tarn.
   (*b*) Suggest a grouping or classification of the lakes shown on the map.
   (*c*) Name the feature between Llyn Padarn and Llyn Peris and, by means of well-labelled diagrams, show how it was formed.

A—Corrie and hanging valley. B—Lateral moraine. C—Medial moraine

PLATE VI A young glaciated valley—Austerdalsbrae Valley, Norway

   (*d*) The Marjelen See in the Bernese Oberland in Switzerland is a type of glacial lake dammed by the Aletsch Glacier. In the School Library you should be able to find a photograph of this lake. Make a sketch of this feature and add explanatory notes.
10. Name three regions outside the British Isles where valley glaciers exist today, and give an actual example of a valley glacier in one of the regions named.
11. Using the O.S. map as an aid, draw contour maps of each of the following: a glaciated valley with hanging valley, waterfall, and cirque.
12. Copy and complete the Table given below, selecting your examples from the Lake District or the Highlands of Scotland.

Features formed by glacial erosion	*Example*	*Features formed by glacial deposition*	*Example*

PLATE VII An old glaciated valley—Nant Ffrancon, North Wales

X to Y—Wide U-shaped valley with receding walls
A—Flat floor, with meandering, misfit stream

13. Fig. 18 shows in profile a corrie, glacier, and the glacier ice feature known as 'icefall'. Copy the diagram and complete the annotations, adding what you consider a suitable title for your sketch.

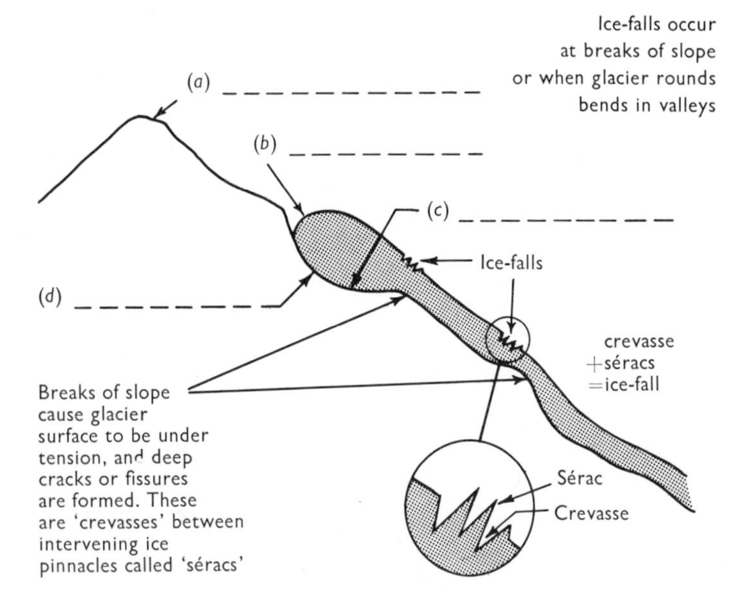

14. Study Plate VII and state:
    (i) the direction in which the camera was pointing;
    (ii) the season of the year when the photograph was taken;
    (iii) possible land use in the area shown in the photograph.
15. (*a*) Settlements are very few and well dispersed; suggest two reasons to explain this.
    (*b*) Make two observations on streams in the photograph area.
16. Examine Map III on p. 33 (1-inch O.S. Sheet 90 Ingleborough area), and you will find in square 7977 Nether Lodge Farm. This is the farm which appears in the centre of the photograph. The drumlins are shown by closed contours. What is
    (i) the average elevation of the photograph area?
    (ii) the average height of the drumlins above the local elevation?

Fig. 18 Glaciated mountain features

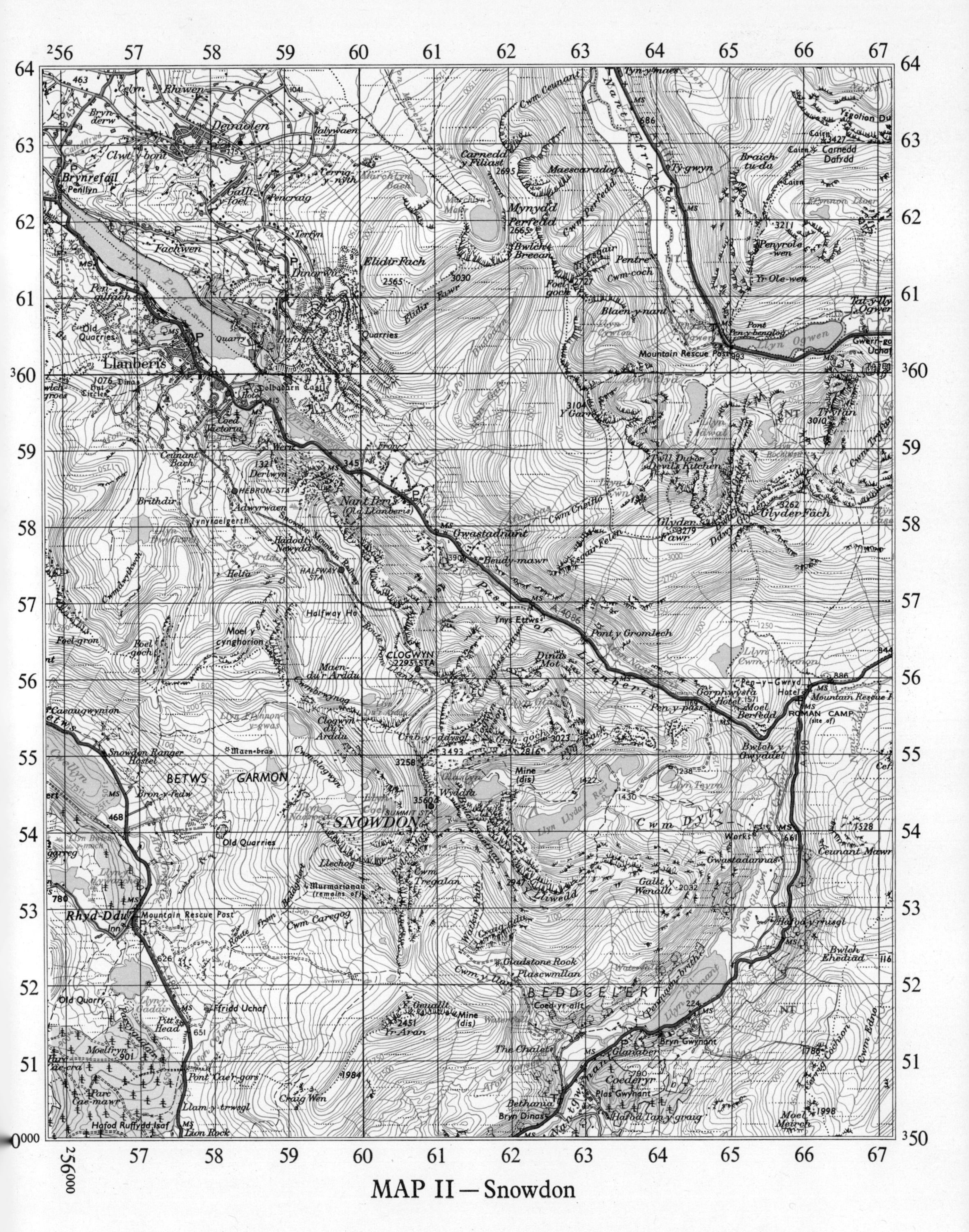

MAP II — Snowdon

nted by Lowe & Brydone (Printers) Ltd., London

5 MOUNTAIN LIMESTONE LANDSCAPES

22 Typical Features of Mountain Limestone Landscape

Limestone consists of calcium carbonate ($CaCO_3$), and is thus known as a *calcareous rock*. It is made from the plant and animal life of the sea, and thus contains many visible fragments of fossilized shells. Limestone is a hard sedimentary rock and forms prominent *escarpments*, especially where it is arranged in horizontal or gently dipping layers or strata.

Limestone dissolves slowly in water containing carbon dioxide (CO_2). Descending rain-water finds its way down along the *joints* and *bedding-planes*, and by dissolving the rock on either side widens them into open fissures. In this way limestone becomes a *pervious rock*, that is, it will allow water to pass through it. Eventually the fissures become so enlarged that most of the rain-water abandons the surface-flow for these newly formed underground channels. Below a certain depth all joints and fissures are permanently filled with water. This depth is known as the *water-table*, and thus becomes the level along which these underground streams flow. Sometimes a surface valley is deep enough for its floor to lie along, or even below, the water-table, so a surface flow is maintained. Other valleys lie just above the water-table, and have no surface flow. They are called *dry valleys* (Fig. 21).

The effect of *ground water* in limestone areas, especially in Mountain (or Carboniferous) Limestone areas, is to produce characteristic and often striking features such as *swallow-* or *sink-holes*, *gorge-like dry valleys*, such as Cheddar Gorge in the Mendip Hills (Plate XI), or Gordale Scar, Malham,

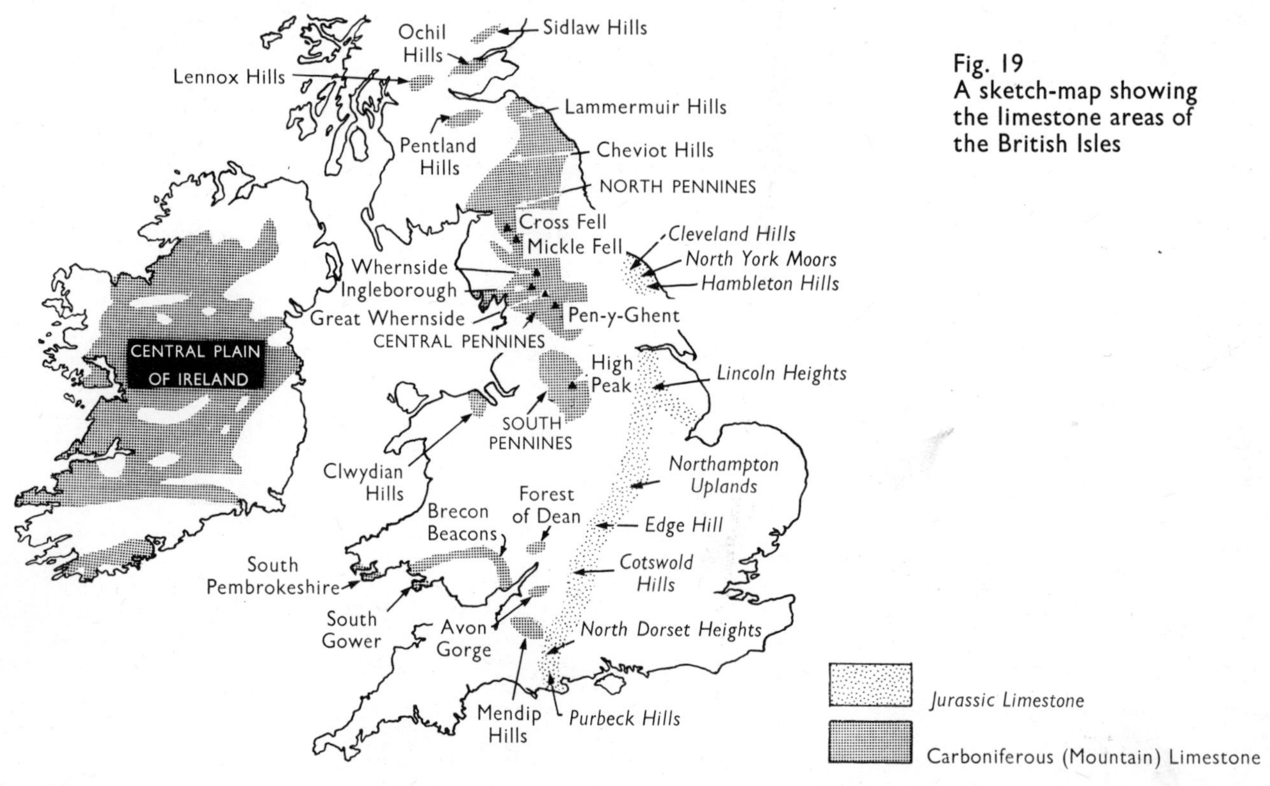

Fig. 19
A sketch-map showing the limestone areas of the British Isles

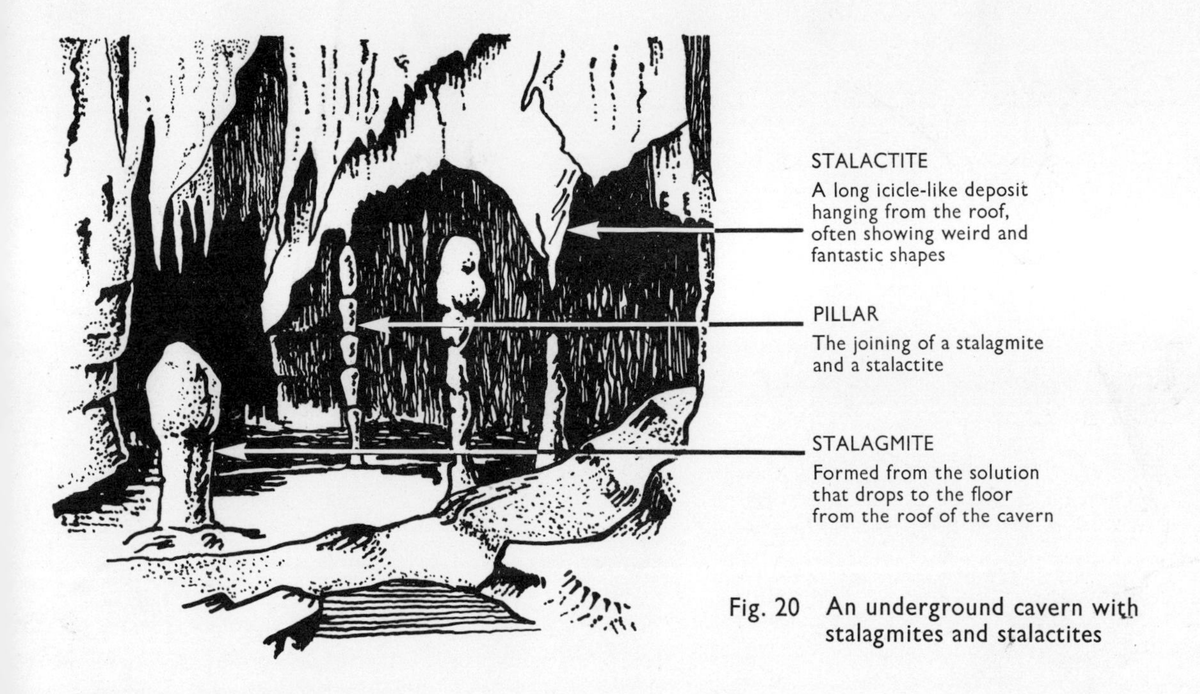

Fig. 20 An underground cavern with stalagmites and stalactites

Yorkshire, and *underground caverns* with deposits of *stalagmites* and *stalactites* (Fig. 20). The landscape of bare rock and thin soils is known as *karst* topography after the Karst region of Yugoslavia and northern Italy. Fig. 19 shows the limestone areas of the British Isles, which cover an appreciable area.

23 The Ingleborough Limestone Area

The Ordnance Survey map extract of Ingleborough (Map III) shows such summits as Ingleborough, Park Fell, Simon Fell, and Whernside (reaching heights of over 2,000 feet), which form part of the Mountain Limestone area of west Yorkshire. Here are seen all the typical 'karst' features. The horizontal 'Great Scar Limestone' is over 600 feet thick. It is 'capped' on most summits by Millstone Grit—a hard, resistant sandstone—and other impervious rocks; these rocks do not readily allow water to pass through them and they thus carry many surface streams. The alternating outcrops of resistant and less resistant rocks give rise to a *step* type of landscape.

These features can be seen on the annotated photograph (Plate XII). There are many bare rock outcrops, and in other places there is only a thin covering of soil. The River Greta flows on the surface for much of its course but, as can be seen on the map extract, there are a few short distances where it disappears underground into the pervious limestone. The main features are illustrated in Fig. 21.

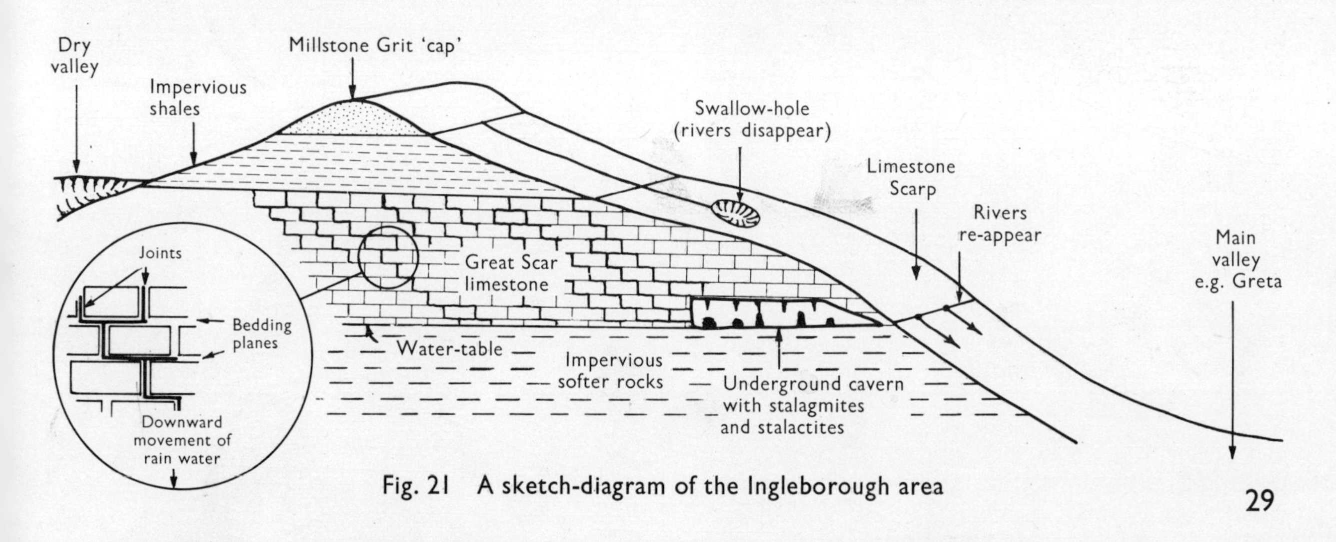

Fig. 21 A sketch-diagram of the Ingleborough area

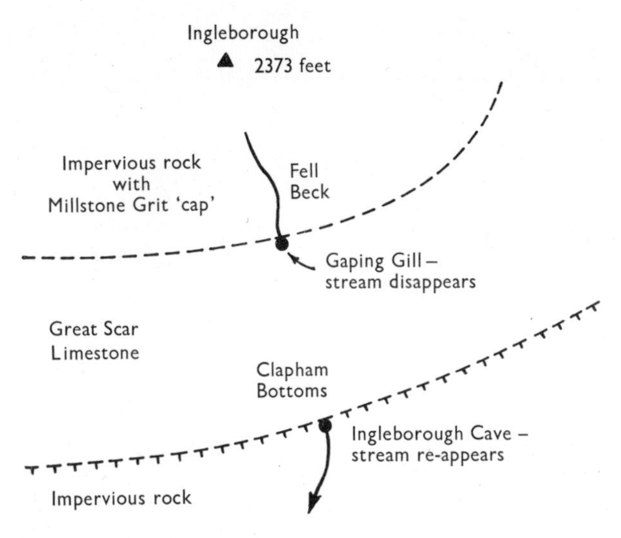

Fig. 22 Stream disappearance at Gaping Gill

PLATE VIII Gaping Gill (or Ghyll), near Ingleborough

Note the disappearance of the Fell Beck at Gaping Gill (Fig. 22 and Plate VIII), south-east of Ingleborough (752727). Another notable pot-hole is Alum Pot (772756)—see Plate IX.

The result of weathering of the bare, jointed limestone is to produce deep grooves, called *grikes*, and furrowed surfaces, called *clints* (Fig. 23), which show clearly the effects of the solution of the pervious limestone by ground water containing carbon dioxide. Clints and grikes together form a *limestone pavement*, as seen in Plate X.

24 Human Activity

Limestone has been quarried for important minerals, such as lead and iron-ore, as well as for lime. Lead mines were intensively worked by the Romans, for example in the Mendip Hills, but they are now largely exhausted, as are most of the Carboniferous iron-ore deposits in the British Isles. Lime is, of course, still used in large quantities in the iron and steel, chemical, and fertilizer industries, whilst limestone is an important building stone and road metal. Because of their thin soil-covering, limestone areas are not much use for arable farming, but support open moorland pasture for sheep.

PLATE IX Alum Pot, on the Horton slopes at Ingleborough

PLATE X Limestone pavement, Ingleborough

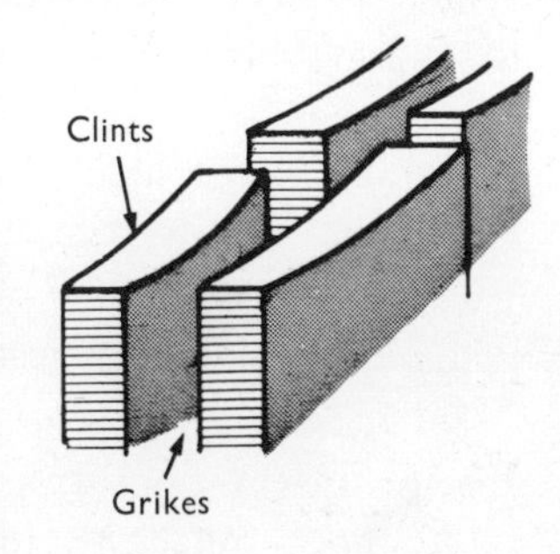

Fig. 23 Clints and grikes

PLATE XI Cheddar Gorge, Mendip Hills, Somerset

EXERCISES

1. Draw a rectangle on half the scale of Ordnance Survey Map III. On your map mark the 1,000-foot and 2,000-foot contour lines, the rivers Greta, Ribble, and Kingdale, and four of the highest summits. What is the area of your map? What is the scale of your map?
2. Name, state the heights of, and give the six-figure map grid-references of the four summits which you have marked on your map.
3. Name the major relief features in squares 7273, 7474, 7572, and 7176.
4. Locate on the photograph, and give the grid reference of: Springcote (note the wooded area); Head Pike; Dale House; West Fell. Calculate the approximate position from which the photograph was taken and state the direction in which the camera was pointing.
5. Draw a section from the summit of Ingleborough (741746) to the summit of Whernside (738813). On your section annotate the following features: River Greta, Roman Road (B6255), Meregill Hole, Chapel-le-Dale. Are the two summits intervisible?
6. In relation to the relief, where would you expect to find
   (i) sheep pasture, (ii) cattle pasture?
7. Find the average gradient
   (i) from Skirwith (707738) to the summit of Ingleborough
   (ii) from Twistleton Hall (702751) to Head Pike (1,886 feet, at 730797).

   Is each pair of points intervisible?
8. Draw a sketch-map to show the disappearance of a stream into Meregill Hole (740758). Mark the heights at which (i) it disappears, and (ii) it reappears on the surface before joining the River Greta. Hence deduce the approximate thickness of the limestone.
9. Relate the routes of the roads B6479 and B6255, and the railway, to the relief.

Head Pike (1,886 feet) with 'cap' of Millstone Grit
West Fell
Whernside (2,419 feet)
White Scars, showing bare outcrops of well-jointed limestone
Rivers disappear here
The Great Scar limestone with 'step' topography
Thin soils with sparse grass cover
Dry walls
Few settlements—sheep-grazing on the hills

PLATE XII Twistleton Scars, from White Scars, overlooking the valley of the River Greta

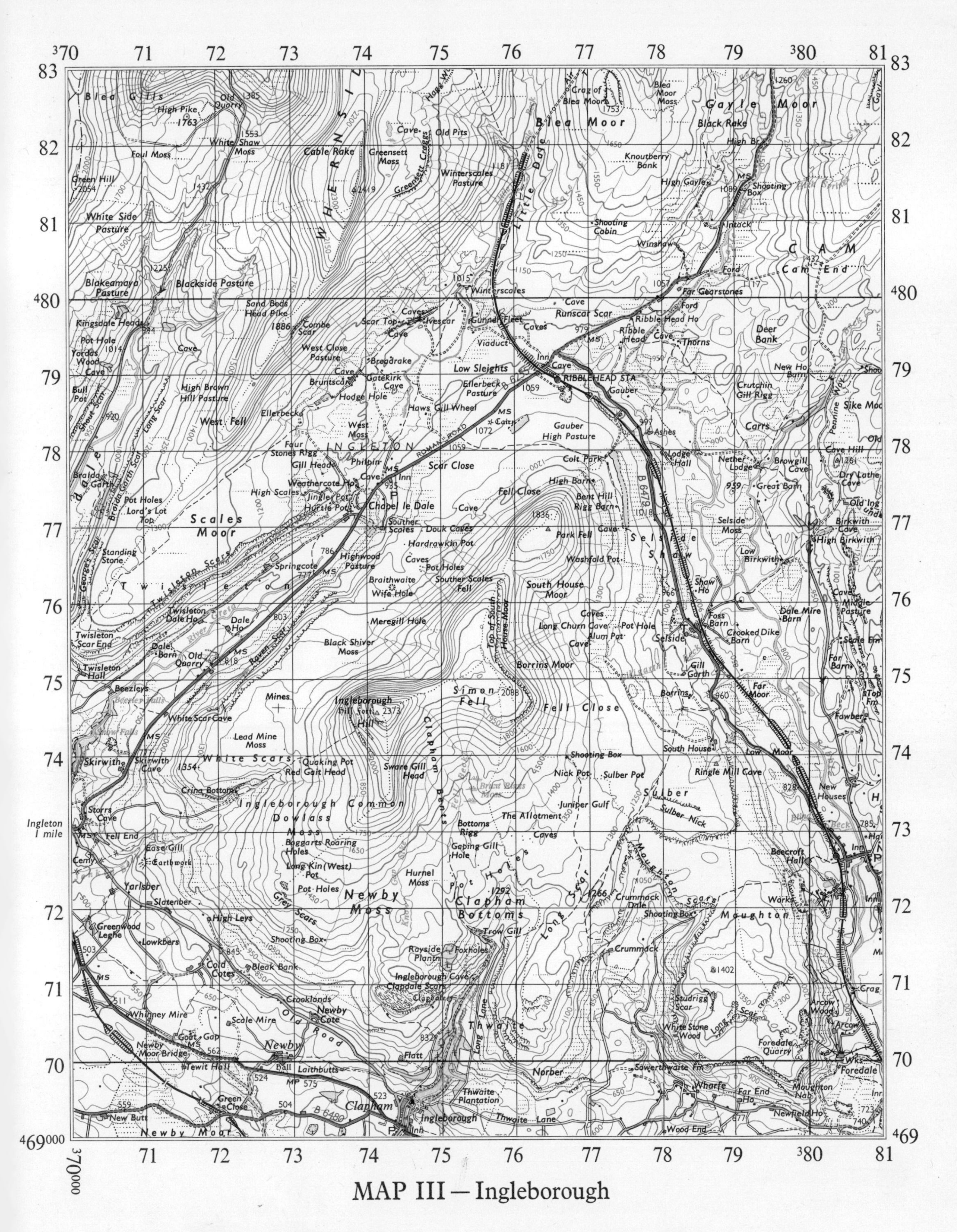

MAP III — Ingleborough

inted by Lowe & Brydone (Printers) Ltd., London

10. (*a*) Describe the distribution of present-day settlement on the map.
(*b*) What evidence is there for saying that the region has been inhabited for a long time? A suggested approach to this question is given below.

11. Describe the stages of the River Greta and its tributaries on the map. Find the average gradient of the River Greta from Winterscales (753801) to west of Skirwith (700738).

12. What evidence is there that the region has been glaciated? Name examples.

Suggested Approach to Question 10

Distribution of Present Settlement

(*a*) Are there 'lines of settlement' on the map, e.g. a spring line, gap line, a valley or route line (following roads, rivers, or railways), a scarp-foot line, an industrial line of mining settlements, or a coastal line of resorts?

(*b*) If so, then briefly explain their occurrence by reference to water supply, shelter, dry site, ease of communications, resources (mineral or farming), and aspect in the case of resorts.

(*c*) Upland areas are usually 'negative zones of settlement' because of isolation, bleakness, wind-swept nature, difficulty of access, and lack of resources.

(*d*) The presence of small towns may be explained by nodality (i.e. a natural meeting-place of routes), a bridging-point, ease of defence (historically), water supply, or a combination of these factors.

(*e*) Clearly the Ingleton map is an example of (*c*) above, where the distribution of settlement is influenced by shelter, water supply, and ease of movement. Thus settlements are found mainly in the valleys of the rivers Greta and Ribble, but even there they consist chiefly of scattered, single-unit farmsteads. The great number of place-names may be confusing, but a study of the map will reveal that most of these are names of relief features. The greatest concentration of settlement is in the south-western corner of the map. There are two reasons for this; state these in your answer.

(*f*) Finally you should note the 'shape' of the settlements. Villages grouped around a focal point (such as a church, village green, castle, bridge, or square) are described as being 'nucleated'. Buildings dispersed along a road give a 'ribbon' shape.

Prehistoric Settlement

This frequently occurs on uplands and is indicated by old-style lettering of names such as Tumuli, Entrenchments, Long Barrow, Celtic Fields. Their positions are well above the marsh and forest which once covered much of the lowland areas of the British Isles. Hill sites were favoured, because they were natural defence points.

6 CHALK LANDSCAPE

25 The Formation of Escarpments in the Wealden Area

Outcrops of chalk, consisting mainly of calcium carbonate, occur in south and eastern England and form distinctive scenery known as *downland*. Fig. 24 shows the distribution of these outcrops, which may take the form of hill ridges like the South Downs, or the broad plateaux of the Salisbury Plain area.

Where the chalk ridges alternate with clay vales the countryside is referred to as scarpland, and here the ridges commonly have unequal slopes—a steep scarp slope and a gentle dip slope. Such a ridge is an *escarpment* or *cuesta*. The geographical features common to scarplands are shown in the sketch transect diagram, Fig. 25. Study these and then identify as many as you can in the accompanying picture of the South Downs, near the Adur River Gap and on the related O.S. 1-inch extract from Sheet 182 (Map IV).

The Setting

Between the escarpments of the North and South Downs is the Weald of Kent and Sussex,[1] an elongated anticline (upfold) composed of strata of

[1] Geologically the Weald comprises the two Wealden Clay Vales and the intervening Hastings Sandstone Ridge. Geographically the Weald is considered to be the region between the two Chalk escarpments.

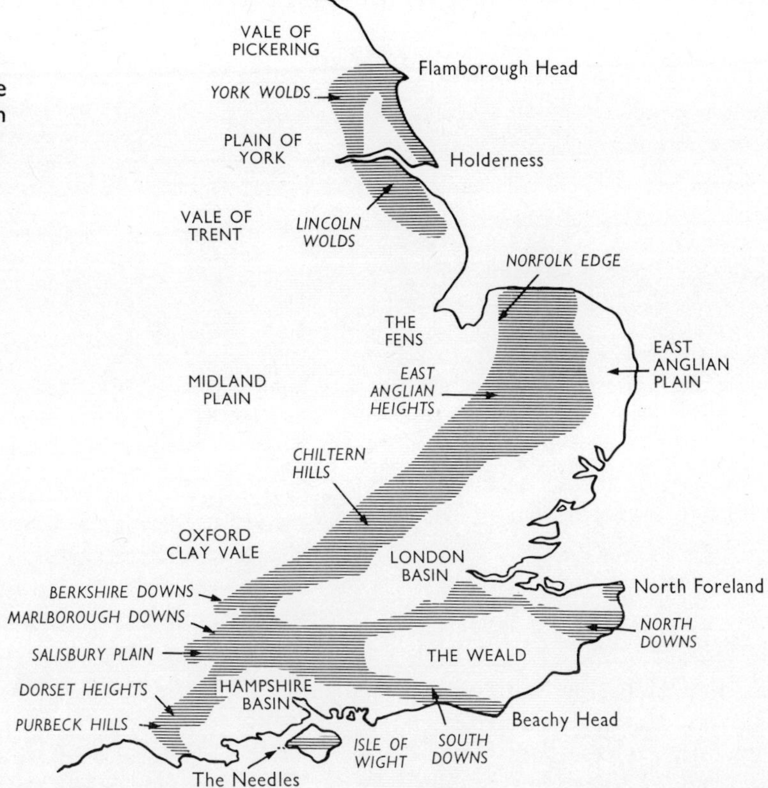

Fig. 24

A sketch-map showing the chalklands of England with clay vales and basins

Fig. 25 A sketch transect, south to north, across the South Downs at the lower Cuckmere valley to show the main geographical features

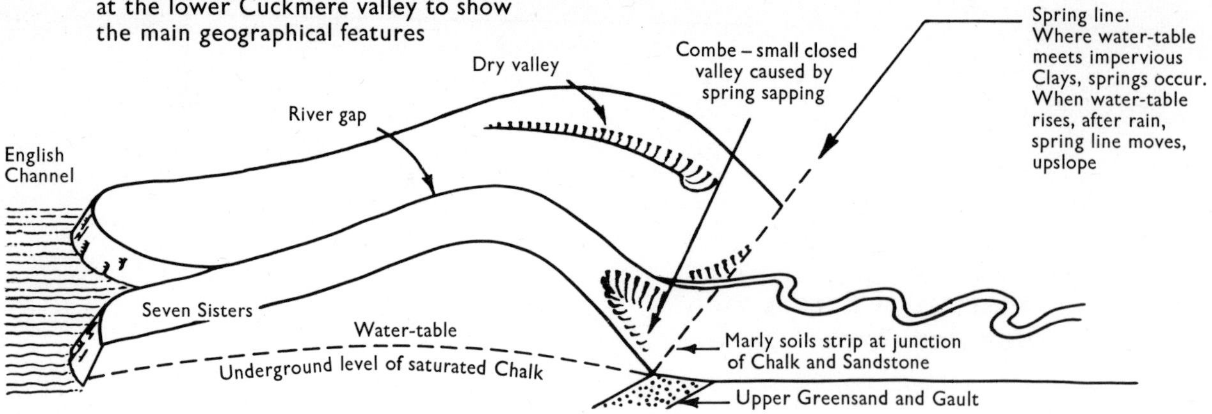

	Feature	*Chalk escarpment*	*Sussex Clay Vale*
1.	RELIEF	Dip slope—many dry valleys. Scarp slope—embayments or coombes	Undulating. Low subdued relief—usually below 100 feet.
2.	DRAINAGE	Absence of surface drainage (Chalk is porous).	Abundance of surface drainage (Clay is impermeable).
3.	VEGETATION	Gorse, thorn bush, downland grasses, with clumps of beech on lower slopes, termed 'hangers'.	Much of natural woodland has been cleared, except for small copses. (Note: the term 'Weald' comes from the German 'Wald'—forest.)
4.	HUMAN FEATURES (*a*) Settlements	No settlements, but much evidence of pre-historic occupation—tumuli, burrows, Celtic fields, camps.	Villages, hamlets and farms on greensand eminences, and on or near the spring line.
	(*b*) Land Use	Sheep grazing. Some cereal cultivation on upper Chalk.	Mixed farming with emphasis on arable activities.
	(*c*) Communications	Farm tracks, 'sunken roadways'. Downs are a barrier to movement, and routeways are restricted to gaps through the Chalk escarpment.	Ease of communications—good network of roads and railways.

varying resistance to the agents of erosion. In consequence, the present relief is a symmetrical arrangement of ridges (of the more resistant Chalk and Lower Greensand) and vales (of the less resistant Gault and Wealden Clays), on either side of a central upland (the Ashdown Forest Ridge of Hastings Sands).

Stages in the Formation of Escarpments

Stage 1. During the Alpine earth movement in Europe in late Tertiary times, the successive strata of Wealden Clay, Lower Greensand, Gault Clay, Upper Greensand, and Chalk were upfolded as shown in Fig. 26. Note that these layers had already been deposited on to the sea floor of Hastings Sands during the Cretaceous period, when seas covered this part of south-eastern England.

Stage 2. With the formation of new slopes, streams known as *consequents* were established (consequents because the direction of flow of these streams was an immediate consequence of the new slopes). Such streams proceeded to remove the chalk cap, already weakened and cracked by the arching of the strata, eventually exposing the Gault Clay beneath.

Stage 3. The more rapid erosion of the clay, aided by the appearance of tributary streams known as *subsequents*, which cut back along the clay outcrop,

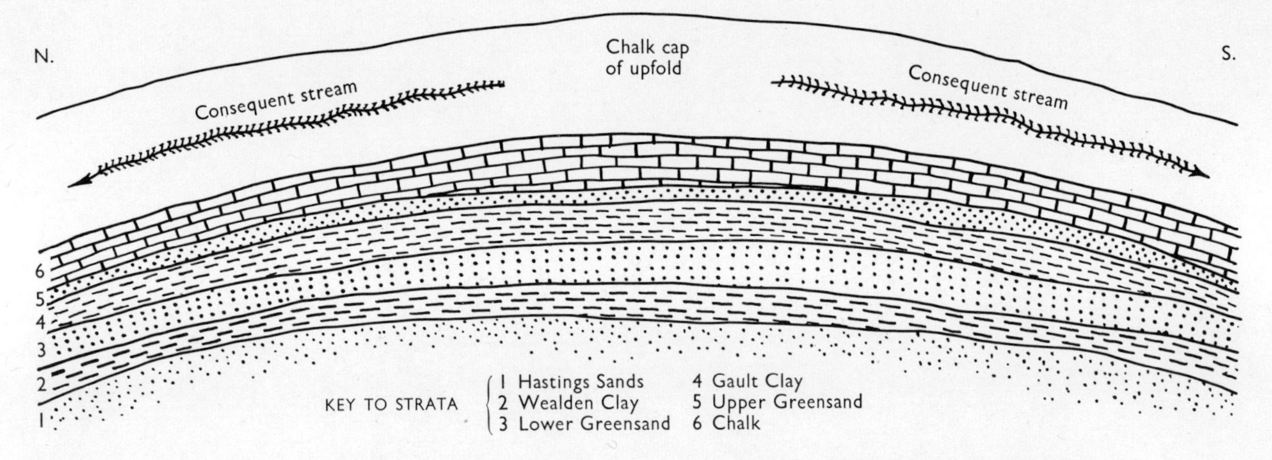

Fig. 26 The upfold of strata in south-eastern England, and the establishing of consequent streams

formed deep valleys overlooked by the steep scarp of the inward facing chalk. These valleys, cut at right angles to the slope of the strata, are called *strike vales*.

Stage 4. Gap cutting through the more resistant chalk, as well as cutting back, i.e. headward erosion, by the consequent streams, proceeded apace. Continued downcutting revealed the Wealden Clay on either side of the resistant Central Ridge, and widening of the clay vales finally produced the present distinctive Wealden relief shown in Fig. 27, in which the escarpments are especially prominent.

26 Features of Wealden Rivers

Most Wealden rivers have their sources in the Central Wealden Ridge of Ashdown Forest. In addition to gap cutting by the northward-flowing Wey, Mole, Darent, Medway, and Stour, and the southward-flowing Adur, Arun, Ouse, and Cuckmere, *river capture* (i.e. the capture of the upper waters of one river by a rapidly backcutting and more powerful neighbouring river) was prominent. There are numerous examples of river capture—such as the Wey beheading the Blackwater to form a wind gap (once occupied by the Blackwater).

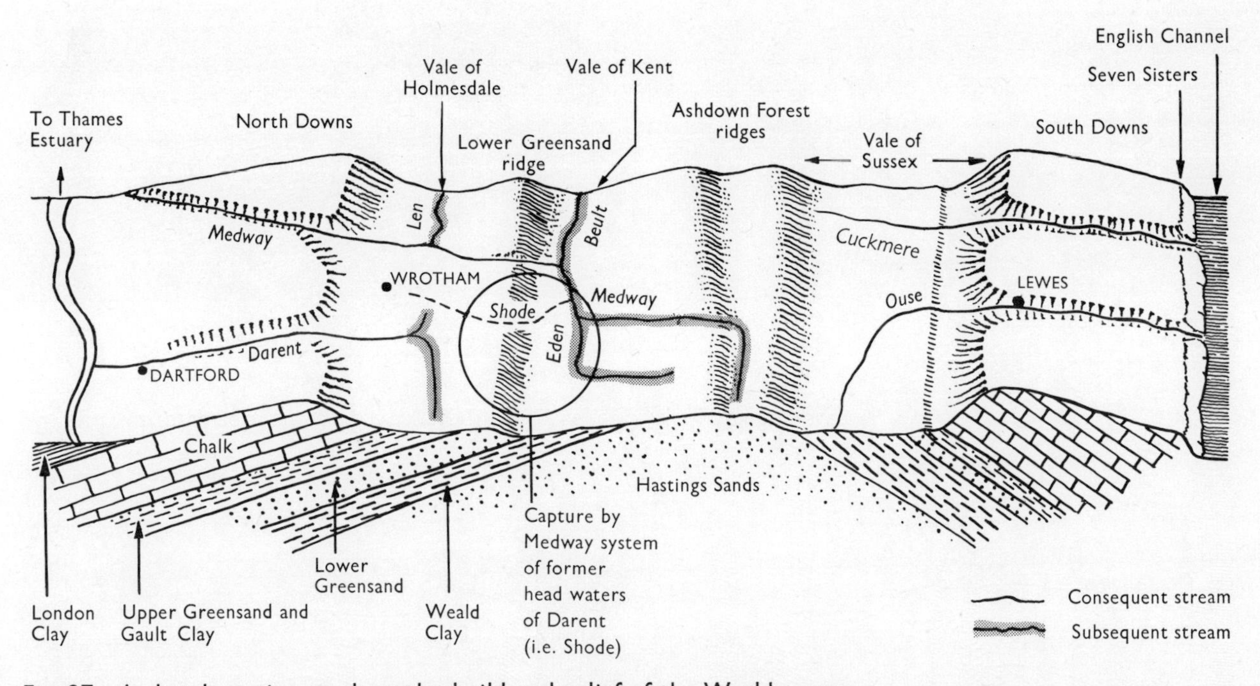

Fig. 27 A sketch-section to show the build and relief of the Wealden area

KEY

X to Y Chalk escarpment

1 Dip slope

2 Scarp slope

3 Wealden clay vale

4 Crest line, joining highest points along summit of scarp

5 Slightly 'sunken' roadways

6 Dry valley—no surface drainage formed when water-table was higher than at present

7 Clay with flint most common on upper chalk

8 Coombe: hollow or small closed valley formed by spring-sapping—the slipping down of saturated chalk

9 Smooth, rounded relief

10 Hanger: beech copse on lower slopes of cuesta

11 Land utilization—arable farming with sheep-grazing on the chalk

A, B, C, D Spring-line settlements. Flint stone used to build churches and old cottages

PLATE XIII

A chalk escarpment, South Downs

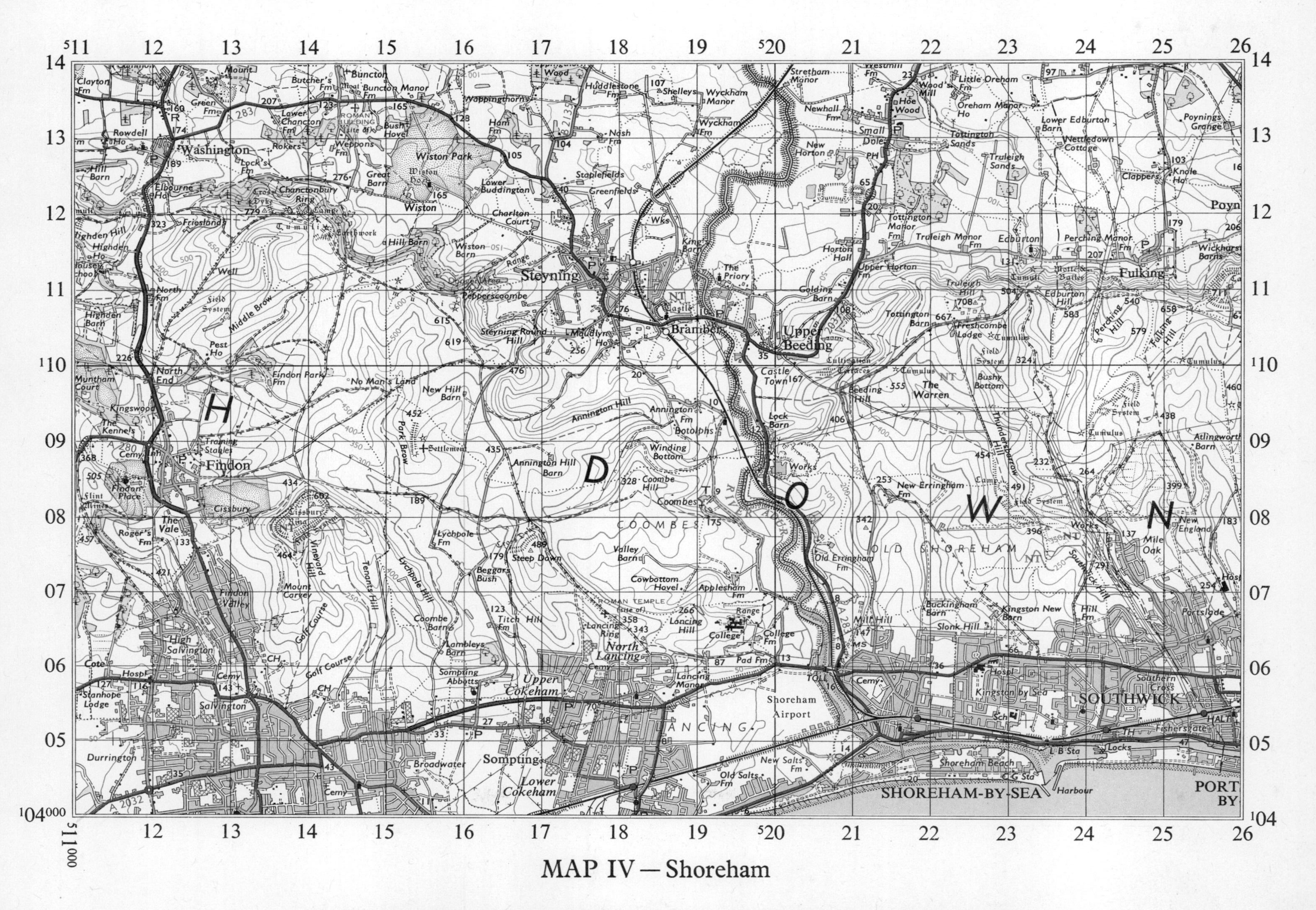

MAP IV — Shoreham

rinted by Lowe & Brydone (Printers) Ltd., London

EXERCISES

Map IV (extract from O.S. 1-inch Sheet 182) and Downs, Plate XIII.

1. Identify the symbols in 2211.
2. Name the two relief features in 2310.
3. Name the two drainage features in 1912.
4. Give the six-figure reference for the upper limit of the tidal section of the eastern branch of the River Adur.
5. What is the chief rock type south of the 12 grid line? List the map evidence in support of your answer.
6. Draw a section from north to south along the 22 grid line between northings 14 and 07, and in tabular form beneath the section insert the relief, drainage, and possible land use along the line of your section. (The section diagram on Fig. 7, p. 11 will help you to answer this question.)
7. Relate the route of the north–south railway to the relief. (A suggested approach to this type of question is given at the end of this series of questions on chalk scenery.)
8. What evidence is there to show occupations in that part of the map area west of the 20 grid line?
9. Draw a sketch-map on half scale to show:
   (*a*) the two main relief regions; insert and number the contour which you have selected as the line of division between your regions;
   (*b*) the highest point, the River Adur, and an example of flood-control drainage;
   (*c*) the two gaps through the upland;
   (*d*) the point from which the photograph (Plate XIII) was taken and the direction in which the camera was pointing.
   (*e*) the important physical feature on the coast.
10. Giving full reasons, state which gives you the clearer impression of the relief of the area—the photograph or the map?
11. Study the photograph (Plate XIII) and answer the following:
    (*a*) The apex of the tree line 'V' near the bottom edge of the picture is at grid reference 253113. Using this as the start-point, measure the approximate length of the scarp slope shown in the photograph.
    (*b*) The letters *A*, *B*, *C*, *D* are printed alongside four settlements; two are villages and two are large farms.
       (i) Name the four settlements.
       (ii) Give two reasons to explain the site of any settlement.
    (*c*) What is the gradient of the scarp slope immediately south of 253113?
    (*d*) How far apart are the two villages, (i) on the map, (ii) on the photograph?
    (*e*) What is the approximate scale of the picture between the two villages?
12. Describe the distribution of vegetation east of the 20 grid line. (A suitable approach to this type of question appears on p. 25, Ex. 12.)

How to Describe Routes in Relation to Relief

(a) **Rail routes.** These are designed to maintain an even gradient and straight course as far as possible. Over undulating ground a level course is achieved by embankments and cuttings. High ground is negotiated via gaps, especially river gaps, or by cuttings leading to tunnels. Valleys are crossed at the narrowest points by bridges or viaducts, which are kept to a minimum. Marshy ground is crossed by embankments.

(b) **Road routes.** The control exercised by relief on road routes is considerably less than in the case of railways, for even steep gradients are no obstacle to motor vehicles. But often upland areas are either skirted, or ascended by a route oblique to the slope, sometimes following a zigzag course by means of a series of hairpin bends. Road routes in the main follow valleys, gaps, dry valley floors, or the lower slopes of uplands adjoining low-lying vales liable to flood.

Notes on the Division of Ordnance Survey Maps into the Chief Relief Regions

(*a*) Difference in general height between one part of the map and another is the basis for dividing the map into its chief relief regions. It would be unusual for more than three chief regions to appear on a map extract.

(*b*) The contour at the foot of prominent slopes—i.e. the obvious break of slope—will be the dividing line between your regions.

(*c*) Number and name the regions on your map. The name should include a reference to its physical nature—vale, plateau, escarpment, etc.—and an indication of its general position, e.g. The Northern Escarpment, Central Vale, etc.

(*d*) Insert the main drainage and the direction of flow.

(*e*) Mark clearly the distinguishing feature in each of your regions, e.g. the flood plain, scarp slope, height of plateau surface, dry valleys, undulating ground.

7 JURASSIC LIMESTONE LANDSCAPES

27 Typical Landscape Features

The Jurassic Limestone escarpments of the British Isles are shown in Figs. 3 and 19. This limestone was deposited in the bed of a shallow sea which covered much of Britain in the Jurassic geological period (see p. viii). It is a rather different limestone from that of the Carboniferous period and was formed about 150 million years ago. Landscape features are not as prominent as those of the Mountain Limestone areas, and summits are appreciably lower, but the escarpment which extends from the Purbeck Hills on the coast of Dorset to the Cleveland Hills in Yorkshire is a distinctive one.

The essential character of these areas depends upon the nature and structure of the rocks. The limestones are *oolitic*, consisting mainly of minute spheres of calcium carbonate, and are full of various types of shells, especially of corals formed during a period of sub-tropical climate. The rock is yellow-brown in colour, forming thin light-brown soils, but often the rock is near the surface and shows bedding-planes, as can be seen in Plate XV, p. 42. In many areas the fields are divided by dry-stone walls, while buildings—cottages, mansions, and churches—are all made of the local stone, called 'freestone'. This can be cut easily when freshly quarried, but hardens on exposure.

28 The Cotswold Area of Gloucestershire

The Cotswold Hills, stretching north-eastwards from near Bath, form a conspicuous part of the Jurassic scarplands. The average height is between 400 feet and 800 feet and rarely does it rise above 1,000 feet. Ascending gradually from the low clay vales of Oxford on the east, the greatest heights are on the western side, where the prominent *scarp* falls sharply to the Vale of Severn.

Plate XIV shows part of the Cotswold scarp near and including *Leckhampton Hill* (965 feet). The photograph follows the line of the scarp, at times very steep with gradients of 1 in 7, and often noticeably concave. Rivers flowing westwards down the scarp begin their courses by having steep gradients and narrow valleys. Note on Map V (extract from Sheet 143, Cheltenham) the course of the stream rising at 800 feet north-east of Dryhill at grid reference 933171.

Plate XV also shows part of the same Cotswold scarp, at the *Devil's Chimney*, Gloucestershire, but this time with the impressive view overlooking the Vale of Severn. Note the two distinctive regions of upland and vale, the wooded area (usually of beech trees), and the bedding-planes of the limestone.

The *dip slope* of the Cotswolds is much less steep

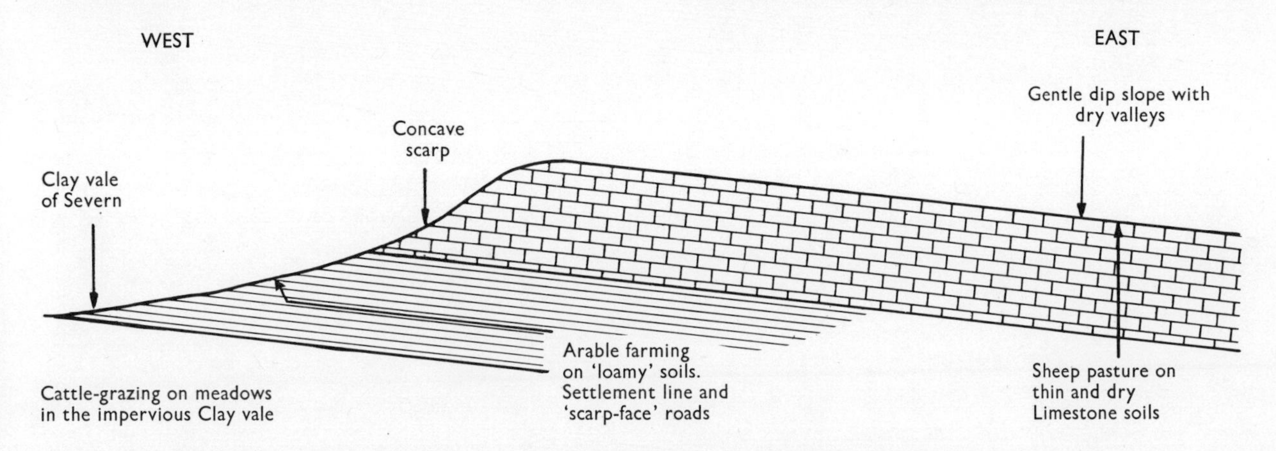

Fig. 28 A sketch-section to show the scarp and dip-slope features of the Cotswolds

PLATE XIV The Cotswold scarp at Leckhampton Hill, Glos.

PLATE XV The Cotswold scarp at the 'Devil's Chimney', Glos.

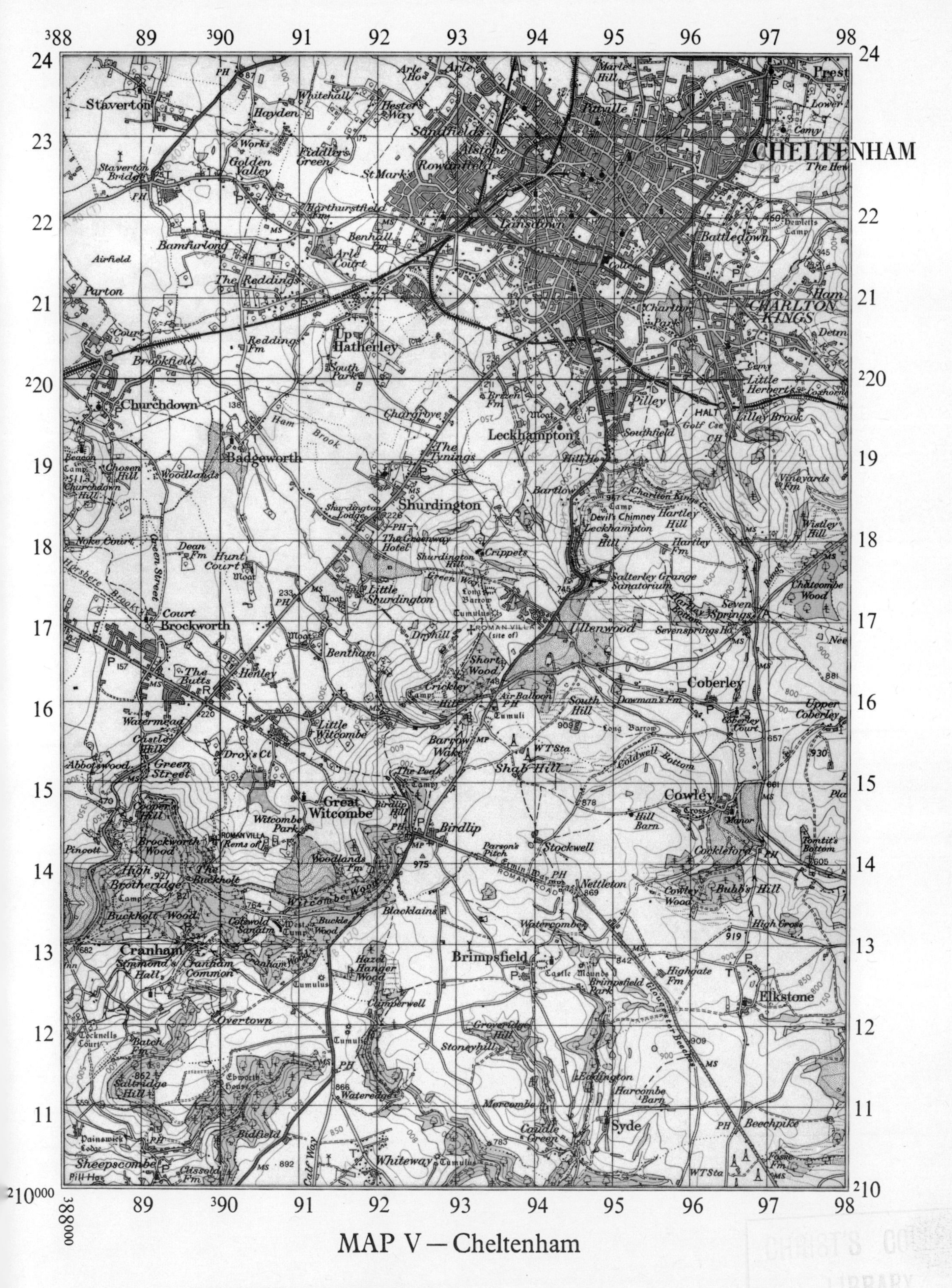

MAP V — Cheltenham

Reproduced from the Ordnance Survey 1 inch map Sheet 143 with the sanction of the Controller, H.M.Stationery Office. Crown copyright reserved.

than the scarp, and the flow of the streams running eastwards towards the Oxford clay vale is much gentler. The most important of these streams is the Thames. Owing to the pervious character of the limestone, many of the valleys are 'dry', but during wet seasons, when the water-table reaches the valley floor, temporary surface streams may flow. The main features can be seen on the annotated sketch-section, Fig. 28.

29 Human Activity

Where the Cotswolds join the clay Vale of Severn a line of *settlements* occurs. These are often market towns, and thus are nodal centres between the adjacent regions. A notable example is Cheltenham, the town in the north of Map V, a focal point of seven Class A roads. Note the route of the road A46 along the lower part of the escarpment. There are no large towns on the upland region, merely a few hamlets and small villages. The woollen industry was important here in the Middle Ages, dependent on the sheep of the uplands and a soapy-clay substance called 'fuller's earth'. This industry has now declined, and a number of the larger settlements specialize in broadcloth, whipcord, and blankets.

Along the border of the limestone upland and the clay vale, crop cultivation is possible on loamy soils, for whereas the limestone soils are light and dry, and suitable only for sheep pasture, the heavier clay soils are largely devoted to cattle rearing and dairy farming.

PLATE XVI

Dragline excavator removing overburden at an open-cast iron-ore mine in the South Lincolnshire Jurassic Limestone belt

30 Other Jurassic Limestone Areas

The remaining areas along the line of the Jurassic escarpment, through the Northampton and Lincoln uplands to the Cleveland Hills of Yorkshire, are much the same as the Cotswold area. There, however, the workings of ironstone become more frequent and economically important, except in Yorkshire, where mines are now closed. Plate XVI shows an iron-ore quarry in south Lincolnshire. Most of Britain's home output of iron-ore is obtained from such open-cast workings. The overburden is stripped by the walking dragline excavator, and when all the ore has been extracted the topsoil is re-spread and the land restored to agricultural use.

In addition to the Cotswold stone, well seen in many fine buildings at Bath, the Jurassic Limestone provides a good building and facing stone in other areas. Examples are the Ancaster stones of Lincolnshire, while the quarries at Barnack in Northamptonshire supply stone used by the building industry at Peterborough and Boston and for many churches in the Fenlands. The limestone at Ketton, in Rutland, supplies local stone, and has been used in some of the Cambridge colleges. Portland stone from the Isle of Portland, Dorset, is a very famous building stone used in many great buildings in London. The Purbeck Limestones are also tough and resistant stones, again widely used for building—for example, in the interior decoration of churches.

EXERCISES

Map V, Cheltenham (extract from O.S. 1-inch Sheet 143).

1. Draw a sketch-map on half the scale of the Ordnance Survey extract and on it mark: the 400-foot contour line; all spot heights and triangulation pillars above 900 feet; the position of Cheltenham, Leckhampton Hill, and all large wooded areas. Find the area of your map in square miles and write down the full Ordnance Survey Sheet number.
2. Give the grid reference of the 984-foot triangulation pillar south-west of Eirdlip, and the triangulation pillar at Leckhampton Hill. Calculate the distance between them in a straight line, in miles and yards, and find whether they are intervisible.
3. Draw a relief section from Reddings Farm (903202) to Hartley Farm (957179), through Leckhampton Hill. On your section annotate: clay vale; scarp; dip; Leckhampton Hill; limestone escarpment; roads, A46 and B4070.
4. Find the gradient from:
   (i) Woodlands Farm (919140) to the top of Birdlip Hill;
   (ii) Leckhampton church with spire to the top of Leckhampton Hill.
5. Describe the physical features in each of the following squares: 9618; 9514; 9418; and 8920.
6. Describe the site and size of the following settlements: Cheltenham; Leckhampton; Churchdown; Shurdington; Syde.
7. Relate the routes of the roads to the relief, and also describe the course of the railway west from Cheltenham.
8. Name any evidence of land utilization shown on the map extract.
9. What indications are there that the map region has been settled for a long time?
10. Write brief notes to explain further the meaning of the following: nodal centre; impervious; 'fuller's earth'; corals; dry walls; loamy soils.

8 RIVERS AND THEIR LANDSCAPES

31 Features of a Normal River Basin

Study the features of a normal river basin shown in the diagram below (Fig. 29). Relate these features to the three photographs, Plates XVII, XIX, and XX, which show the source area, a well-formed river valley, and a river-mouth area respectively. Note how the form of the river valley changes while its width increases with the general lowering of the surrounding land. The changes represent different stages in the life history of a *normal* river—that is, one which develops in a region such as the British Isles, with all-year rainfall, where the processes of weathering combine with the river and its tributaries to form the landscape.

32 Stages of River Development

The three stages referred to above are recognized as *youth*, *maturity*, and *old age*.

(a) *Youth*

This stage is illustrated by Plate XVII. It begins when a stream establishes itself in an area affected by mountain-building—that is, on a newly uplifted slope—and begins to cut for itself a course down-slope. The average gradient of the young stream is steep and its course has many *rapids* and *waterfalls*. Pot-holes, drilled by swirling stones, eventually join, thus lowering the stream bed. Such downcutting or vertical corrasion is the main work of the youthful stream, and enables it to cut a deep, narrow, steep-sided valley with many *interlocking spurs*. Silt and other eroded material carried downstream, especially during flood, becomes the *load* carried in suspension.

(b) *Maturity*

In this stage, the lower parts of interlocking spurs are removed by the increasingly swinging action of the river—now greater in volume, having been joined by *tributaries*—to form well-marked *bluffs*. Maturity is indicated by the stage when the valley floor has been widened to accommodate the increased flow from tributaries into the main river. *Lateral erosion* or valley-widening is now the chief work of the river. In this stage the gradient is more regular, as many of the rapids and falls have been removed to form a smoother profile (the long-section of a river's course from source to mouth) than was the case in youth. Across the valley floor a *flood plain* is formed by deposition of some of the river's load. Note that the *meander zone* is approximately the same width as the flood plain itself, which is bounded by prominent bluffs. Some of the features associated with old age may also appear in this mature stage.

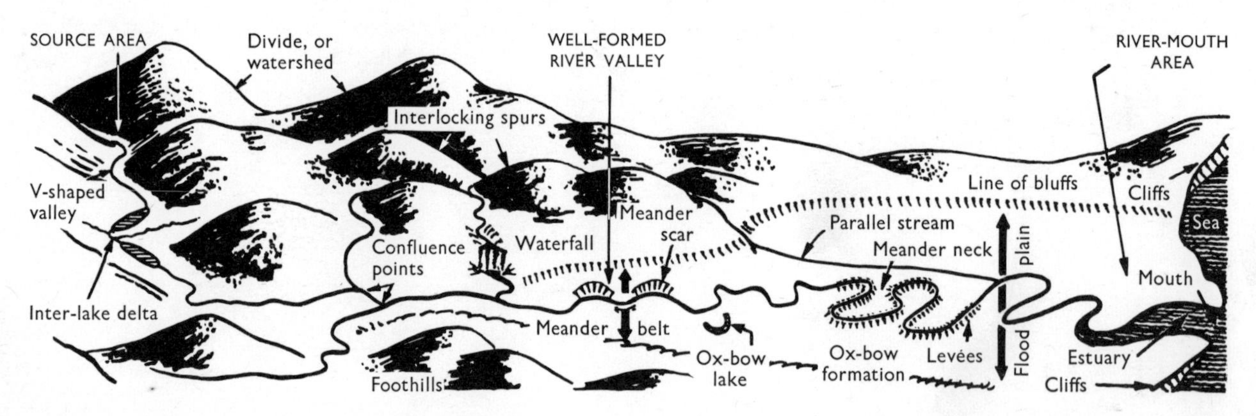

Fig. 29 Features of a normal river basin

(c) *Old Age*

Old Age begins when the bluffs are lowered and pushed back, so that the flood plain becomes many times wider than the meander zone. Deposition is the main work—forming a *raised bed* and natural embankments, extending the flood plain, and under certain conditions assisting the formation of *deltas* and *distributaries*. *Ox-bow lakes*, *parallel streams*, and *delayed confluences* are also characteristic of a river in old age. The whole landscape is very subdued, the gradient slight, and the profile of the river's course very low and smooth.

It is important to appreciate that no river in the British Isles is in old age from its source to its mouth. Many rivers in highland Britain exhibit all three stages in their course, and in others the development of old age has been interrupted by uplift of a localized nature causing *rejuvenation*, i.e. the cutting afresh of its bed in old age. Such downcutting, really a feature of youth, causes the *meanders* to become *incised* into the flood plain. That section of the flood plain then becomes a prominent *river terrace* above the new, lower level of flow. The break in the previously smooth profile of the river is known as the *knick point*.

33 The Appearance and Mode of Formation of some River Features

(a) *Pot-holes*

Pot-holes are a feature of a river in youth. They are drilled in the source area or mountain tract of a stream bed by stones and pebbles swirled round by the flow of water. This rotating action eventually forms a deep smooth-sided pot-hole as well as wearing down the pebbles. When the pot-holes join, the bed of the river is appreciably lowered and this is the method whereby a river downcuts its bed and deepens its valley. *Example:* the Strid section of the River Wharfe in the Yorkshire Dales.

(b) *Rapids and Waterfalls*

Rapids are stretches of a river, usually in youth, where the flow is swifter than in the stretches above and below them. Rapids are generally formed where hard and soft rock beds outcrop in the river's course. Erosion of the soft, less resistant beds leaves ledges, over which the river flows more rapidly. Extensive and steep rapids are called *cataracts*.

At well-marked breaks of slope, where the more resistant strata lie horizontally over softer ones, the

PLATE XVII The youthful valley of Crossdale Beck, Cumberland

PLATE XVIII Thornton Force, near Ingleton, Yorks.

eroding of the latter causes the river to fall over the hard rock ledge, forming a waterfall. *Example:* waterfall and rapids at Thornton Force, near Ingleton, Yorkshire (Plate XVIII).

(c) *Meanders*

These are the sweeping bends or curves in a river's course. Meandering, which becomes evident in the mature stage of a river's development, is the most prominent feature of a river in old age, as seen on Plate XX, but it is important to appreciate that the formation of meanders is a continuous process, which begins in youth and continues through to extreme old age. The actual cause of meandering is thought to be the greater eroding power of the river's current on the outside bend, as shown in the diagrams (Fig. 30). Continuous erosion on an outside bend undercuts the bank to form a *meander scar* or *river cliff*, while on an inside bend some deposition will take place, because the current is slacker here and a sloping low spur or *slip-off slope* is formed. Prolonged erosion of meander scars will cause the meander to shift downstream to form the *meander neck* marked on Plate XX. Flood water cutting through the meander neck will in turn form the isolated river arc —the *ox-bow lake.* Study Fig. 30, which illustrates the formation of meanders.

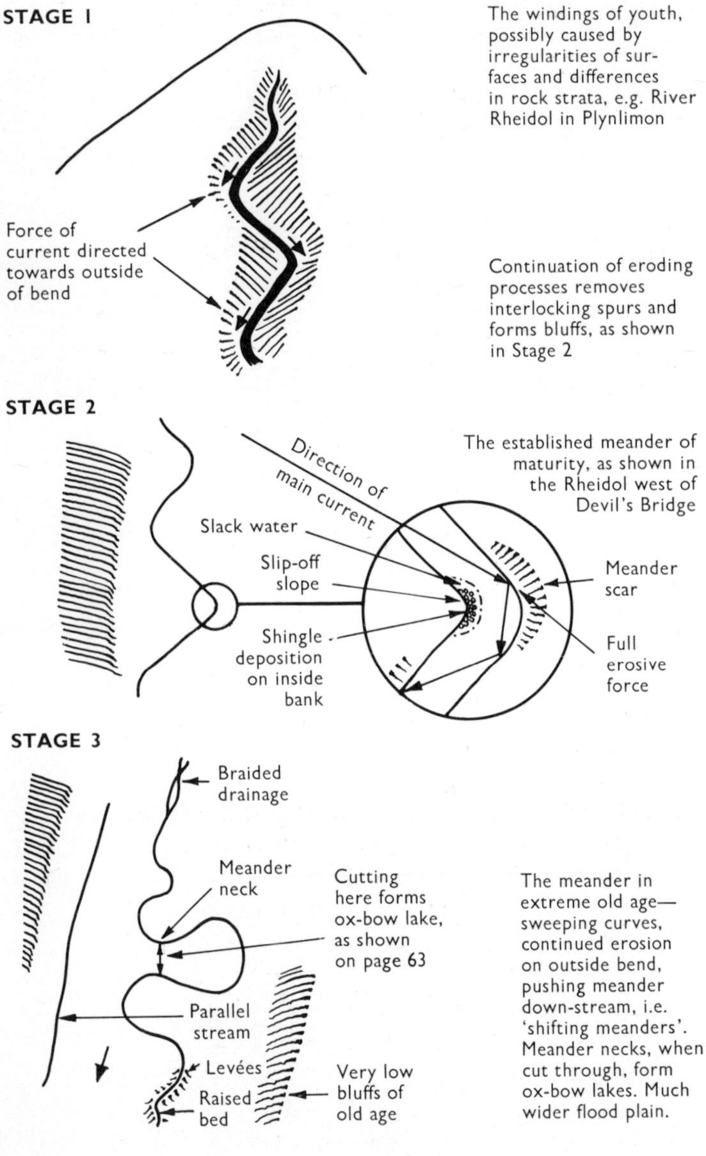

Fig. 30 The formation of meanders

(d) *Deltas*

Deltas are extensive flats of mud and silt formed by deposition at points in the river's course where the speed of the current is checked, e.g.

(i) where a mountain stream strikes the valley floor, the sudden check to its flow forms an *alluvial fan* or *miniature delta.*[1]

(ii) where a river enters a lake, the still water checks the flow to form a *lake-head delta.* Where two rivers enter the lake on opposite sides a delta may link up with another from the far shore to form an *inter-lake delta,* such as that separating Buttermere and Crummock Water in the Lake District.

(iii) where a river enters the sea a delta will form wherever deposition of alluvium is in excess of the amount removed by sea currents and tides.

There are no large open-sea delta formations in the British Isles (can you suggest reasons for this?), but the build-up of alluvium is rapid in enclosed seas such as the Mediterranean and Caspian, and also in partially enclosed seas like the Gulf of Mexico. In the sea areas the saline content is high, and this assists the more rapid precipitation of the finer silts carried by the river, and the consequent build-up of the delta. Continuous deposition raises the general level of the delta and causes it to extend seawards. As the delta level approaches that of the river, the river breaks into a number of channels known as *distributaries,* on the banks of which, during periods of flood, low embankments of silt are formed. Behind them the appearance of marsh and swamp vegetation indicates the final stage in the conversion of what was once a sea area into a new land area, which will continue to grow seawards.

[1] Alluvial fan is the term given to a fan-shaped deposition area of silt on the land floor of a valley. If deposition is on the lake floor, it is described as a miniature delta.

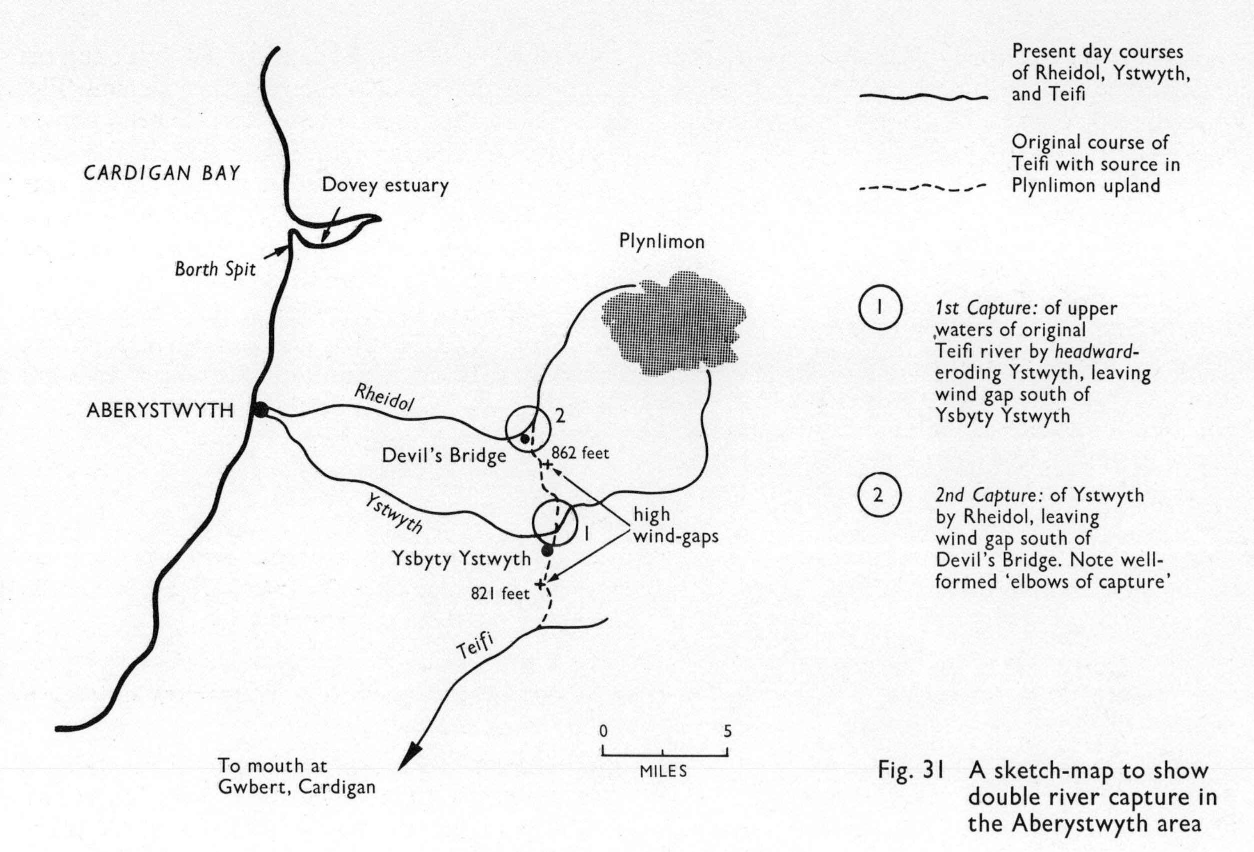

Fig. 31 A sketch-map to show double river capture in the Aberystwyth area

Note: Though the term *delta* comes from the Greek capital letter delta (Δ), not all deltas are so shaped. In addition to the symmetrical *fan* shape of the Nile, there is the *bird's foot* type of the Mississippi, the *flat arrowhead* type of the Tiber, and the *estuary* type of the Seine.

An example of a distributary is the Petit Rhône which flows out of the Grand Rhône near Arles, in the lower Rhône Valley.

(e) *River Capture*

The capture of the headwaters of one river by another is a feature of the drainage of the Weald in south-east England, and of south Wales. It occurs when the greater *headward erosion* (erosion backwards from the original source) of the *pirate river* cuts into the course of the invaded river. The resulting diversion of flow dismembers the latter river, which now becomes a *beheaded stream*, and that section of the river's course between the point of capture and the new source of the beheaded stream becomes a *wind gap*, or a high dry valley. A fine example of double river capture is provided by the Teifi-Rheidol-Ystwyth group of rivers in west central Wales. Fig. 31 shows the sequence of river capture; note especially the two *elbows of capture*.

EXERCISES

Preliminary Note

When you are requested to describe the course of a river, ensure that the following points are dealt with in your description:

(*a*) Whether the selected river is the main drainage system on the map.

(*b*) The direction of flow—this can be determined by noting the heights and the shape of contours which cross the river valley; the apex of the V of the contour points towards the source. Where tributaries join the main stream, the apex of the V of the confluence points towards the mouth. Spot heights along the river banks may also help you to determine the direction of flow.

(*c*) The stage of development attained by the river—youth, maturity, or old age. Very rarely will a river show all three on a single map extract. Describe the features which are associated with the stage named and which are clearly shown on the map.

(*d*) Any special features—natural, such as river capture or incised meanders; or man-made, such as flood-control drainage.

KEY

X to Y The flood plain is approximately the same width as the meander zone

1 The limits of the flood plain are marked by bluffs

2 Abandoned meander ('ox-bow' lake)

PLATE XIX The mature valley of the River Glass, Strathglass, Inverness-shire

Map VI, Strathglass (extract from O.S. 1-inch Sheet 27).

1. The bridge which appears in the immediate foreground in Plate XIX is the northernmost of the two shown on the map. Give the six-figure map reference for the bridge.
2. State the direction in which the camera was pointing.
3. Complete the following: 'The camera was pointing towards the —— of the river.'
4. With reference to the photograph only, describe the direction of flow of the River Glass. Give one reason to support your answer.
5. Write a brief reasoned account of the distribution of vegetation shown in the photograph (Plate XIX).
6. A close study of the photograph shows that on the plateau-like surface of the upland there are several light-coloured patches. What are these? Suggest one reason for their occurrence.
7. Describe the stage of the River Glass shown on the map.
8. Draw a sketch-map (half-scale) of the area east of the 40 grid line and insert the following: (i) a confluence point, (ii) the main watershed, (iii) stream braiding, (iv) mixed woodland, (v) the name and height of the highest point, (vi) the two main areas of settlement.
9. Give three good reasons to explain why settlement is so sparse west of the River Glass.
10. Show how relief has influenced communication in the map area.

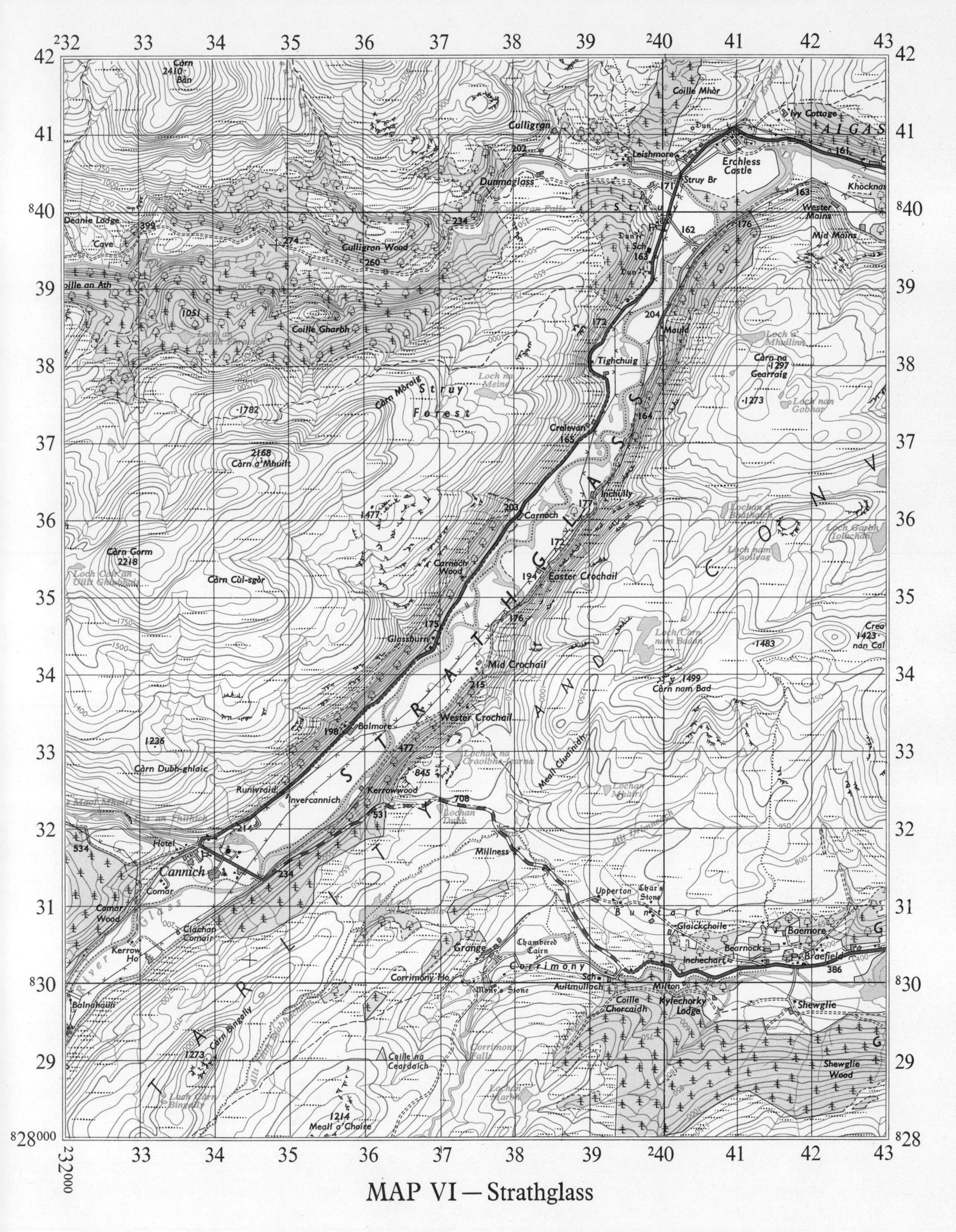

MAP VI — Strathglass

Reproduced from the Ordnance Survey 1 inch map Sheet 27 with the sanction of the Controller, H.M.Stationery Office. Crown copyright reserved.

Printed by Lowe & Brydone (Printers) Ltd., London

PLATE XX A river in old age—the Forth at Alloa, near Stirling

X to Y Limits of meander zone
1 Meander neck
2 Slip-off slope
3 Meander scar
4 Braiding
5 Alloa

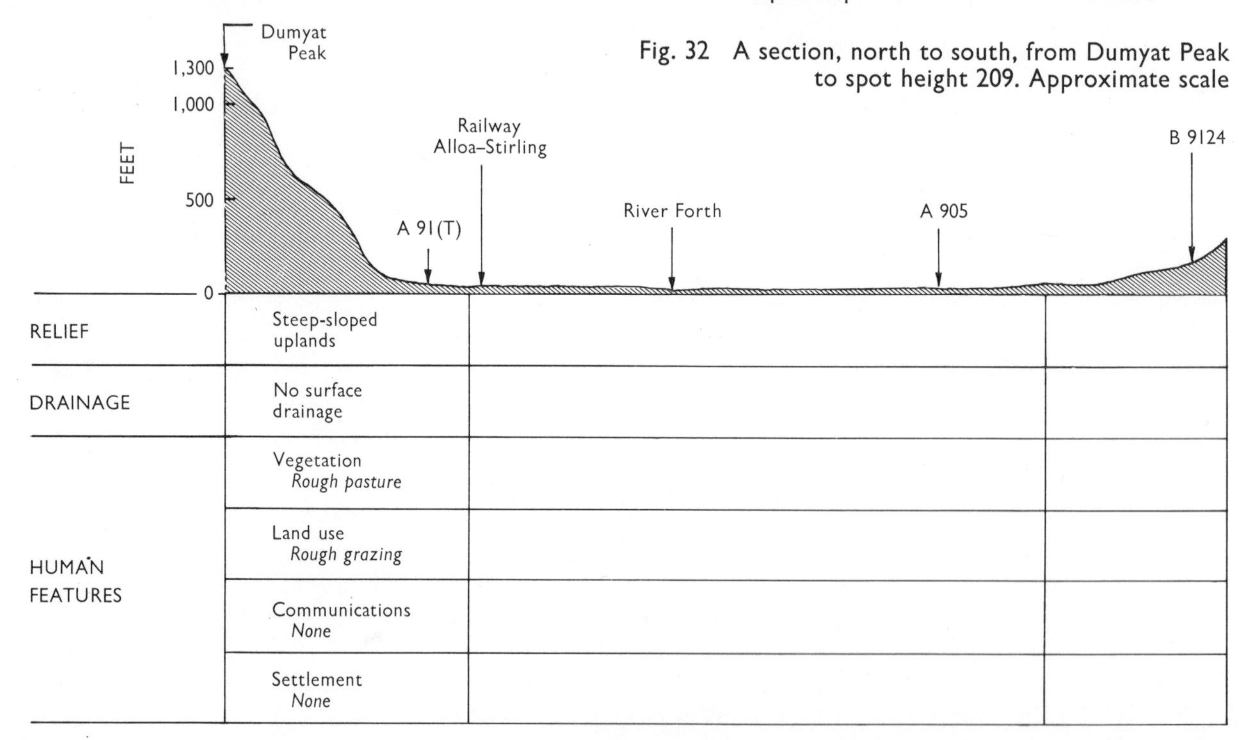

Fig. 32 A section, north to south, from Dumyat Peak to spot height 209. Approximate scale

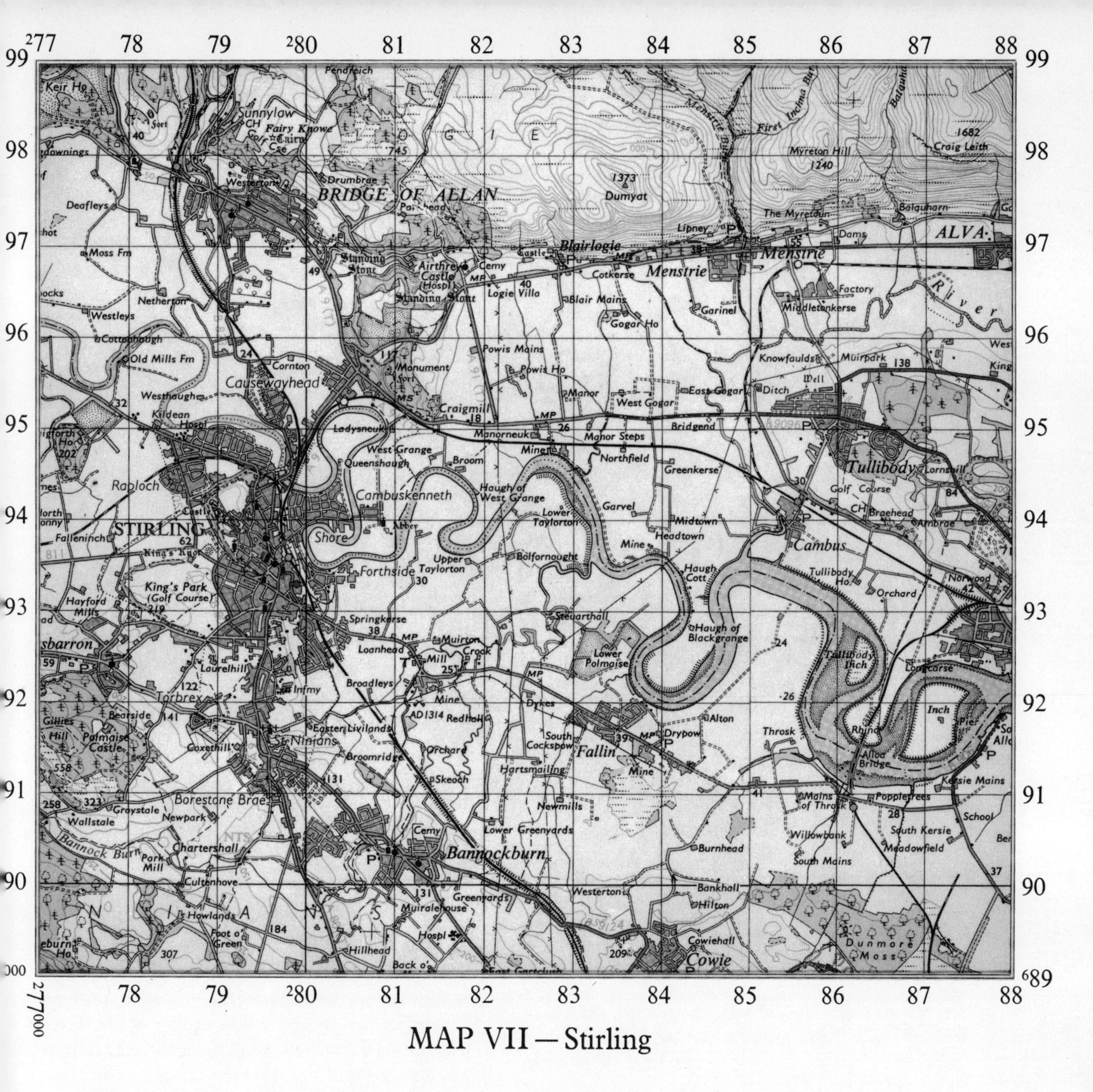

MAP VII — Stirling

Reproduced from the Ordnance Survey 1 inch map Sheets 54 and 55 with the sanction of the Controller, H.M. Stationery Office. Crown copyright reserved.

Map VII, Stirling (extract from O.S. 1-inch Sheets 54 and 55).

11. What is the gradient between the mile-post on A91(T) (8396)—estimate its height—and Dumyat Peak (8397) to the north of the mile-post?

12. (*a*) What is the straight-line distance between the mine at 827947 and Lower Taylorton also in 8294, (i) on the map, (ii) on the photograph (Plate XX)?

    (*b*) Calculate the approximate scale of that section of the photograph (Plate XX).

13. Draw a sketch-map to show what the map tells you of (i) the site of Stirling and (ii) its function.

14. (*a*) Draw a section from Dumyat Peak (8397) to spot height 209 in 8389.

    (*b*) Under your section draw and complete a table similar to the one shown opposite. Select the necessary information from the map strip between grid lines 83 and 84.

The combined section and table in Fig. 32 is known as a *transect*, that is, a section showing the physical characteristics and human features of the area covered by the section.

The photograph shows evidence of both the destructive and the constructive work of the sea. The more resistant basalt forms prominent headlands, while the more easily eroded chalk forms wide bays

In front of each headland is a wave-cut platform—the base rock remnant of that part of the headland eroded by the sea

The plateau-like land area is also thought to be a product of sea erosion—the 'planing' of the coastal area by the sea during an earlier geological period, when this land area was submerged

PLATE XXI

Whitepark Bay, County Antrim

9 COASTAL SCENERY

34 The Causes of the Variety in British Coastal Scenery

Nowhere is coastal scenery more varied than in the British Isles. With the exception of the coral coasts of tropical regions, most types occur, reflecting very closely the different rock structures of Britain.

The variety of coastal forms is due to the twin processes of wave action—erosion and deposition—acting on rocks of different hardness; the more resistant rocks form prominent *headlands*, while the more easily eroded rocks give rise to wide *bays*, as shown on Plate XXI.

35 Sea Erosion—the Destructive Work of the Sea

Wind-formed waves are the chief agents of sea erosion.

Longshore drift is responsible for the transporting and depositing of eroded material.

The sea erodes by:

(*a*) the shattering pressure of pounding waves;
(*b*) the battering effect of waves armed with pebbles and rock fragments;
(*c*) the undercutting of a cliff face by the rapid expansion and contraction of air in the crevices into which water is forced;
(*d*) the dissolving of limestone rocks in salt water.

The effect of the four processes is the eventual removal of headlands and the filling-in of bays, resulting in the general straightening of the coast. The features formed by the work of the sea on headlands are shown in Fig. 33. Plate XXII of Old Harry Rocks, Dorset, shows many of the features referred to in the diagram.

The rate of cliff retreat depends on both the hardness of the rock and the dip of the strata, as shown in Fig. 34.

Sea erosion is most severe in Yorkshire, Lincolnshire, and Norfolk. For example, during a storm in February 1953 part of the Lincolnshire coast—of boulder clay and glacial sands—retreated some 50 feet. The normal rate of retreat is about 6 feet per year. Where a hard rock overlies a less resistant stratum, erosion of the latter causes slipping and subsidence, as at Lyme Regis (Dorset), where the chalk cliffs have slipped over the clay base.

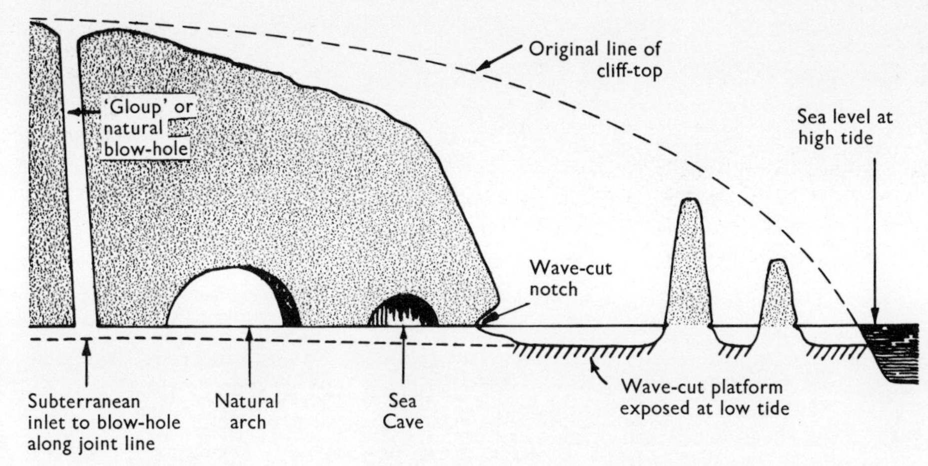

Fig. 33 Coastal features formed by sea erosion

STAGES IN THE REDUCTION OF A HEADLAND

1. Cutting of 'wave-cut notch' in cliff face
2. Formation of caves by various processes of sea erosion – aided by jointing of cliff rocks
3. Link-up of caves on both sides of headland to form natural arch
4. Enlargement of arch – collapse of arch roof to form stacks
5. Removal of stacks, and levelling of wave-cut platform

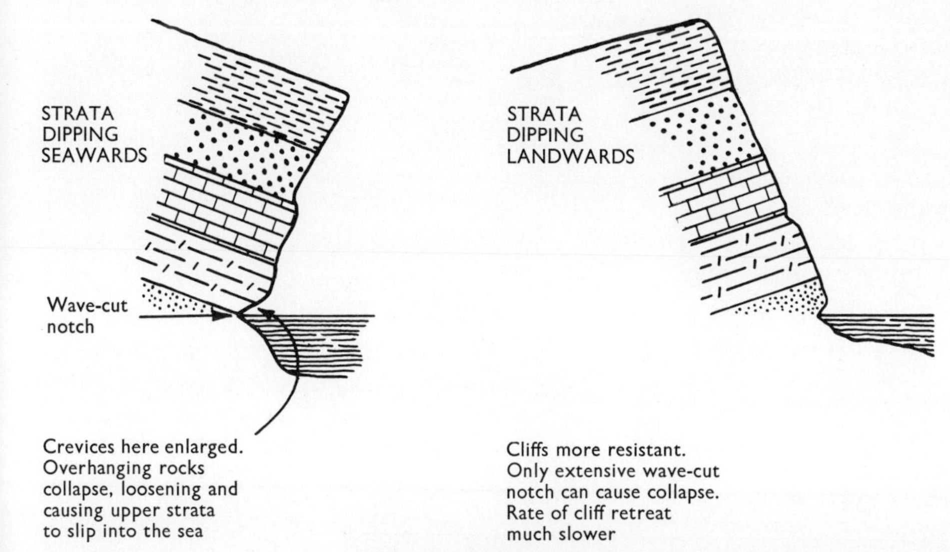

Fig. 34 How sea erosion is assisted or checked by the dip of coastal features

PLATE XXII

Stacks at Old Harry Rocks, near Studland, Dorset

36 Deposition—the Constructive Work of the Sea

Material eroded from headlands is deposited in bays and low-lying coasts by (i) wave and tidal action, and (ii) *longshore drift*. Study Fig. 35 and you will understand how longshore drift both transports and deposits eroded material.

Features formed by the two processes (i) and (ii) above are:

(*a*) **Beaches.** A beach is that part of the shore which lies between the highest spring tide and the lowest neap tide marks, and on which the waves break. It may consist of mud, sand, gravel, shingle, or pebbles. During gales, banks of pebbles may be thrown up to form *storm beaches*, as at Pevensey Bay and Cuckmere Haven in East Sussex.

(*b*) **Sand dunes.** These are formed when beach sand is dried and piled up by the wind on the land fringe of beaches. Extensive sand dunes occur near the mouths of large estuaries and around wide bays, as at Pendine Sands, Carmarthen Bay.

(*c*) **Bars and Spits.** Bars are offshore banks of wave-transported beach-type material deposited on the sea floor. Bars may be slightly above or below sea-level, and those with one end joined to the land and the other in open water are known as *spits*. Plate XXIII shows how spits are formed.

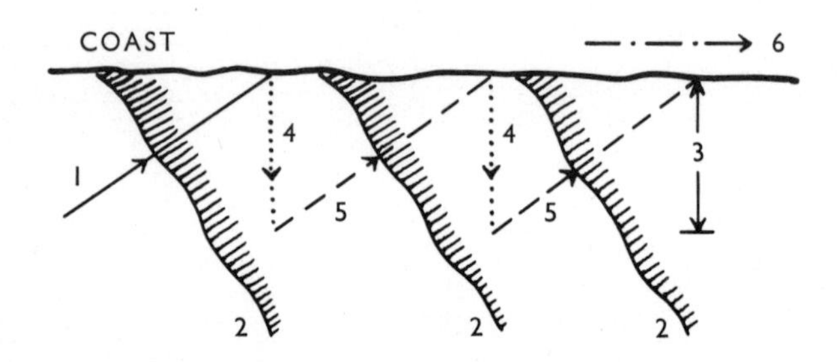

1. Direction of wind, and of movement of advancing waves
2. Waves break here
3. Depth of beach exposed during backwash
4. Direction of backwash, carrying beach material with it
5. The same material carried obliquely forward by next wave
6. Direction of longshore drift

Fig. 35 The action of longshore drift

37 Types of Coasts in the British Isles

Since the end of the Ice Age the coasts of the British Isles have been partially submerged by a very gradual rise in the level of the sea. This encroachment by the sea followed the melting of the glaciers and ice sheets, which once covered the British Isles north of a line joining Bristol and London. The most extensive of the ice caps was centred over the Highlands of Scotland—where its weight caused a sagging of the earth's crust in that region. The removal of the ice cap enabled parts of Scotland to rise at a more rapid rate than that of the level of the sea. Hence while parts of the coasts of the British Isles exhibit characteristics associated with submergence, others owe their distinctive features to emergence from the sea. This distinction is accepted as a basis for the classification of British coasts.

(a) *Submerged Type of Coast*

First we shall consider three examples of submerged coasts:

(i) Ria Coast. This comprises a series of drowned river valleys following the submergence of an upland coast, forming long parallel inlets separated by elongated peninsulas, as in the case of the Dingle, Kenmare, and Bantry Bays in south-west Ireland. Elsewhere, as in the south-west peninsula of England, the rias are branching in form, as in the Kingsbridge estuary and in Plymouth Sound with the Lynher, Tamar, Tavy, and Plym rivers. Fig. 36 illustrates the main features of ria coasts.

(ii) Fiord coast. Fiords are partly drowned glaciated valleys, and exhibit many of the features of such valleys—the U-shaped cross-section, the parallel steep walls in that section of the original valley occupied by glaciers, and the hanging tributary valleys. It is important to note that partial submergence was inherent in their formation, in that the glaciers, owing to their tremendous thickness and weight, scoured parts of the valley floor below sea-level. Erosion continued on to the sea floor itself until the glaciers were converted into icebergs. Elsewhere melting may have occurred at the mouth of the valley. This would account for the shallow entrances to some fiords, through which sea water could enter the fiord and submerge the lower sections.

In the British Isles, fiords, though not identical in all respects to those in Norway, are found

PLATE XXIII

A raised beach, on the west coast of Great Cumbrae Island, Bute

along the west coast of Scotland. They are often referred to as *sea lochs* and fine examples are Loch Leven and the adjoining Loch Linnhe (Fig. 37). Study Figs. 36 and 37, and Plates XXIV and XXV, and note the prominent differences between ria and fiord coasts.

(iii) Lowland estuary coast. This is the counterpart of the ria coast, but in lowland areas. Such a coast is typical of the Essex coast, where the estuaries of the Rivers Thames, Crouch, Colne, and Stour are the partly drowned flood plains of these rivers. These estuaries have branching inlets but no inter-estuary peninsulas.

(b) *Emerged Type of Coast*

The **raised beach** is the best example, and occurs widely in the British Isles—especially in Scotland along both the sea lochs of the west coast and the firths of the east coast. A feature of the raised beach is the presence of beach deposits on this level sea-cut bench. The photograph of the west coast of the Great Cumbrae Island, Bute (Plate XXIII), shows clearly how the sea-level has fallen, and the extent to which the sea has retreated from the old cliff line to expose the raised beach, along which runs a coastal road.

PLATE XXIV A ria—the Dart estuary, South Devon

Features of a Ria Coast

1 Gently sloping sides—an open V-shape—where hilly regions are submerged. But slopes are steeper if the region is a submerged dissected plateau
2 From the mouth to the ria head the water becomes shallower
3 The beach is exposed at low tide
4 Ridges between the valleys form peninsulas, surrounded by the branching inlets of the ria
5 A sheltered harbour: Plymouth is used as a naval base, Milford Haven as an oil terminal

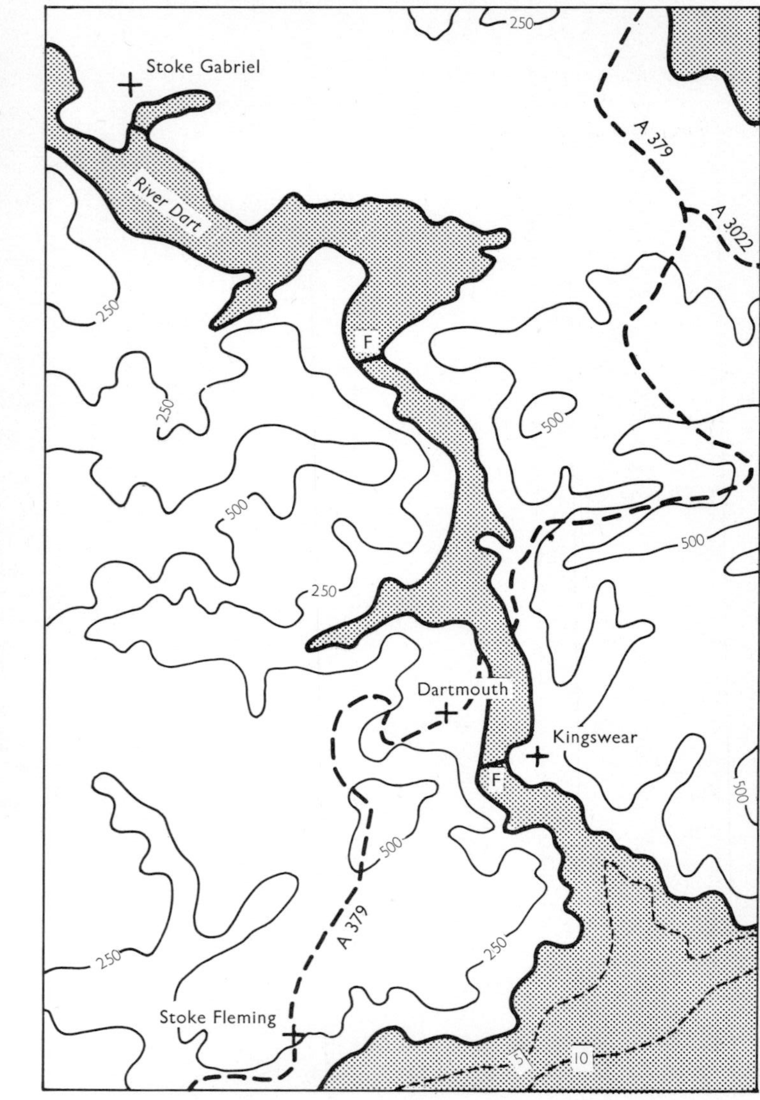

Scale: 1/63,360

Contours at 250 ft.-intervals

F = Ferry

Fig. 36 The Dart estuary—a ria

PLATE XXV

A fiord—Loch Leven, Argyllshire

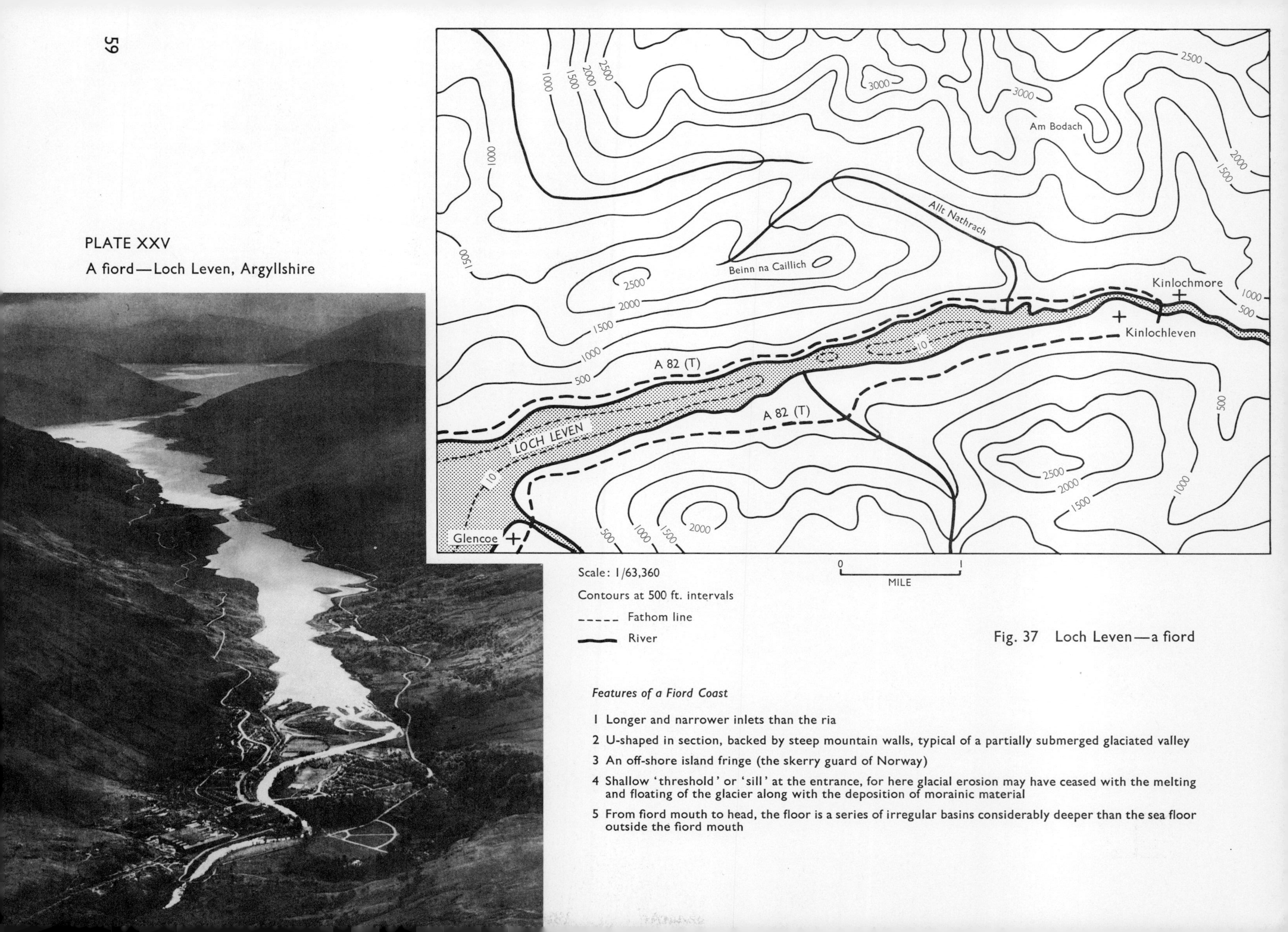

Scale: 1/63,360

Contours at 500 ft. intervals

Fathom line

River

Fig. 37 Loch Leven—a fiord

Features of a Fiord Coast

1 Longer and narrower inlets than the ria

2 U-shaped in section, backed by steep mountain walls, typical of a partially submerged glaciated valley

3 An off-shore island fringe (the skerry guard of Norway)

4 Shallow 'threshold' or 'sill' at the entrance, for here glacial erosion may have ceased with the melting and floating of the glacier along with the deposition of morainic material

5 From fiord mouth to head, the floor is a series of irregular basins considerably deeper than the sea floor outside the fiord mouth

PLATE XXVI Ro Wen Spit, south of Barmouth

The formation of the Ro Wen Spit

1 Upland coast—source of supply of sand and pebbles produced by erosion and transported to build up the spit

2 Direction of longshore drift, the agent of transportation

3 Direction of prevailing wind-formed waves

4 At estuary the change of direction of coast, and check to longshore drift by River Mawddach, enable the waves to pile drift material to form the spit

EXERCISES

(Map VIII, Barmouth (extract from O.S. 2½-inch Sheet SH 61) and Ro Wen Spit, Plate XXVI.

1. Name the coastal features represented by the symbols at (i) 608149, (ii) 623148.
2. Distinguish between the two black-coloured linear symbols in square 6114.
3. (*a*) Give the map reference of the point of entry on the map of the River Mawddach.
   (*b*) How wide at low tide is the River Mawddach at this point?
   (*c*) How wide at high water (ordinary tides) is the estuary of the River Mawddach along a line through this point?
4. (*a*) What is HWMOT in 6214?
   (*b*) Calculate the area of sandy beach exposed at low water between grid line 62 and the railway in 6214.
5. Draw a sketch-map of the land below 50 feet south of the River Mawddach and clearly mark five types of land use. Add a key, and indicate the approximate scale.
6. Describe in relation to the relief the route followed by the railway from Fairbourne Station to the station in Barmouth.
7. What is the general upper limit to woodland in the map area? Give two other facts relating to the distribution of woodland in the map area.
8. With the aid of the photograph and the map draw an annotated sketch-map to show how the Ro Wen Spit was formed.
9. Give the approximate position of the camera, and state the direction in which it was pointing when the photograph was taken.
10. Draw a sketch-map to show as much as you can of the geography of Barmouth.
11. Determine whether or not the school 637141 is visible from the Garn in 6116.
12. (*a*) What is the relief feature at 624140?
    (*b*) What are saltings in 6415?
    (*c*) What is Fegla Fawr at 6214?
13. Draw a sketch-map to show the main geographical features of the map area.
14. Excluding the estuary, there are three relief regions in the map area; name and locate these regions.
15. Under the headings relief, drainage, land use, settlements and communications, list the main contrasts between the Barmouth map area and your school area.

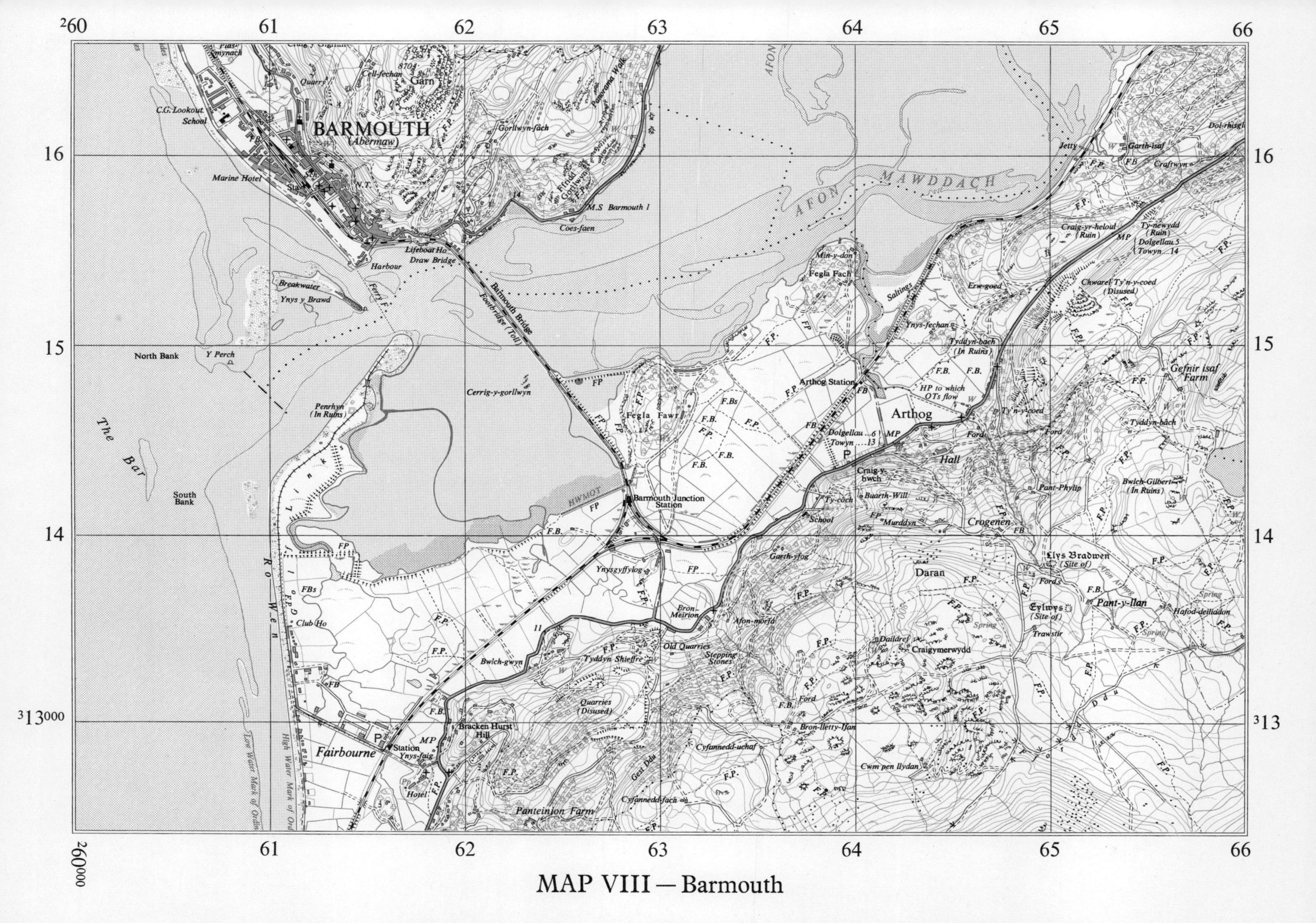

MAP VIII — Barmouth

nted by Lowe & Brydone (Printers) Ltd., London

10 LAKE SCENERY

38 Definition of a Lake

A lake is a hollow or depression on the earth's surface, filled with water. Lakes vary in size—from the ponds, meres, and tarns so common in many parts of the British Isles to large inland seas, such as Lake Superior and the Caspian Sea, in other parts of the world.

Most lakes are formed naturally, but some are man-made (or at least enlarged by man), by building a dam across the valley so as to form a reservoir capable of supplying water to a large town or city. For example, Lake Vyrnwy in North Wales is used by Liverpool in this way.

39 Common Lake Types in the British Isles

More lakes are due to the action of glaciers than to any other mode of formation. Numerous examples are found in the highland areas of the British Isles.

The photograph of Thirlmere (Plate XXXI) in the Lake District shows an example of a *finger lake.* This type of lake is formed in an over-deepened U-shaped glaciated valley. It is long, and in places very deep, and is 'dammed' by a terminal moraine which forms an effective barrier; hence it is sometimes called a *barrier lake.* When the ice disappeared the valley was partially blocked by the moraine barrier, and water accumulated behind it (Fig. 38*a* and *b*).

A tarn (Fig. 39) is a small glacial lake which occupies the floor of a cirque (see Chapter 4).

The photograph (Plate XXVII) shows a tarn (Glaslyn) in Snowdonia, and the cirque or cwm in which it is contained. There are a number of examples of tarns to be seen on Sheet 107 (Snowdon).

The third type of glacial lake is a *kettle lake.* It occupies a hollow in a glacial drift area and is formed by the melting of a large block of ice which became isolated from a glacier. Kettle lakes are usually small, and are numerous in glacial deposition areas. A

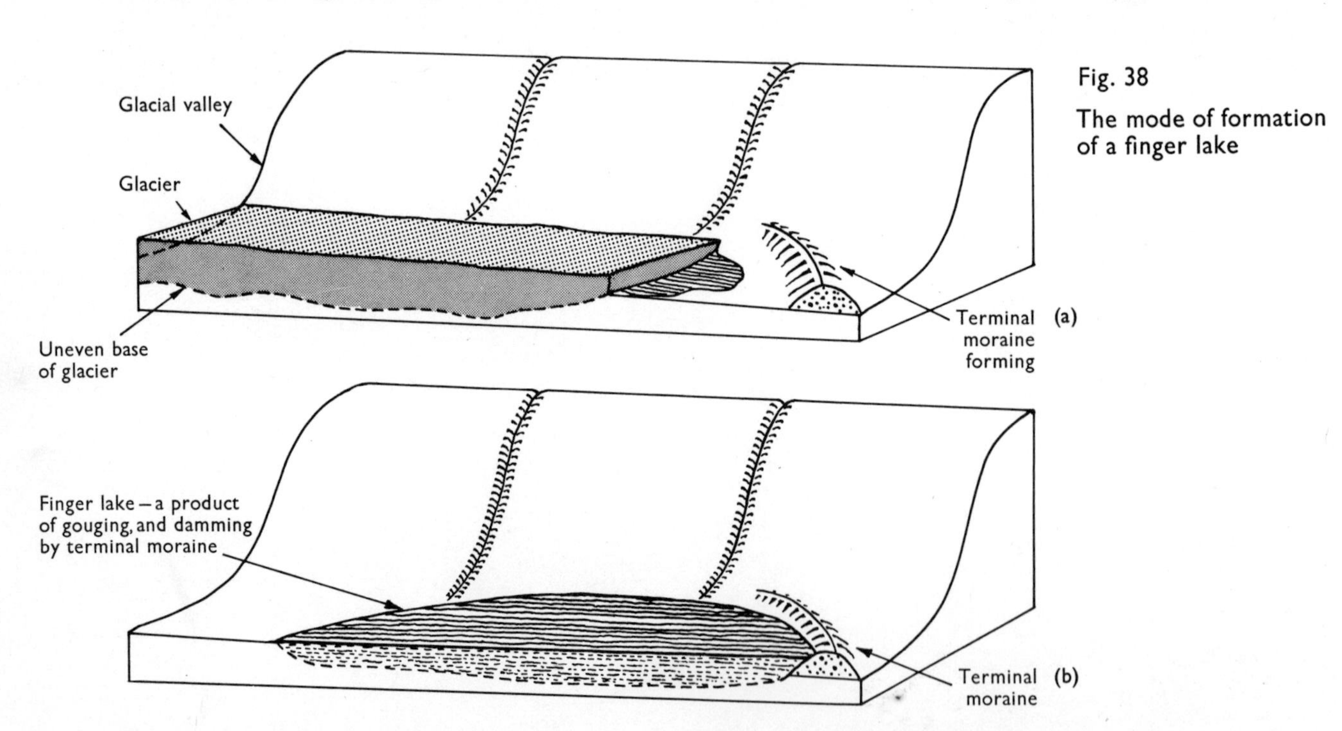

Fig. 38
The mode of formation of a finger lake

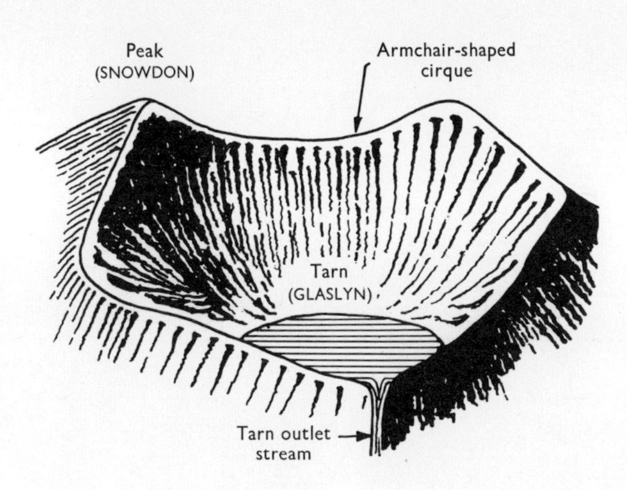

Fig. 39 A cirque and tarn

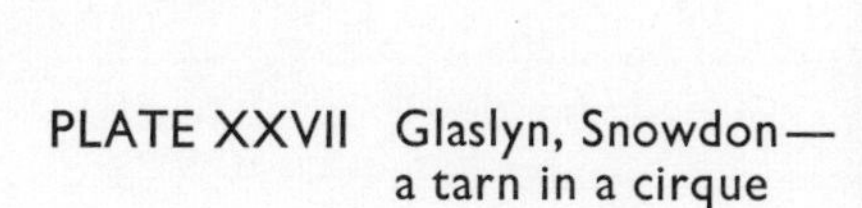

PLATE XXVII Glaslyn, Snowdon—a tarn in a cirque

typical example (Plate XXVIII) is found near Cornhill, Northumberland, in the Tweed Valley (note also the 'drumlin' topography), while the Strathglass Valley (Plate XIX) also has a number of 'kettles'.

An *ox-bow* (*cut-off* or *mortlake*) is formed by a river, usually in its plain stage. It is the crescent- or horseshoe-shaped lake made by the river when it cuts through a meander neck. These abandoned meander loops are then sealed off at both ends by silting (Fig. 40).

An ox-bow lake appears in Plate XIX. See if you can identify it and give its grid reference on Map VI, Ordnance Survey Sheet 27.

Limestone *solution lakes* are caused by the solution of the soluble limestone resulting from the flow of a river over this type of rock. Such lakes are common in Central Ireland along the River Shannon, an example being Lough Derg. This lake is over 20 miles long and below it are the falls of Killaloe, which supply water power for the hydro-electric power plant near Limerick, about 12 miles away.

Cheshire is an important salt-working area of Britain. Solution and subsidence of an impervious surface have led to the formation of a *mere* (or *flash*) after the salt has been pumped out. This type of solution lake is usually quite shallow.

Fig. 40 The formation of an ox-bow lake

PLATE XXVIII A kettle lake at Cornhill, Northumberland

CLASSIFICATION OF LAKES

Type	*Cause*	*Kind*	*Example*
1. EROSION	(*a*) Glaciation	(i) Tarn	Llyn Idwal, North Wales
		(ii) Finger	Thirlmere, Lake District
		(iii) Kettle	Cornhill, Northumberland
	(*b*) Solution	(i) Mere	Cheshire
		(ii) Limestone solution	Lough Derg, Ireland
2. DEPOSITION	(*a*) Glaciation	Moraine-dammed (same as finger lake above)	Thirlmere
	(*b*) River	Ox-bow	Strathglass Valley
	(*c*) Coastal	Lagoon	Loe Bar, Helston, Cornwall
3. EARTH MOVEMENT	Large 'hollows' in the earth's crust, for example a rift valley	(i) Depression	Lough Neagh, Northern Ireland
		(ii) Rift valley	No direct example in the British Isles. (Dead Sea, Red Sea, Lake Malawi, Lake Tanzania)
4. VOLCANIC	Craters of extinct volcanoes	Crater lakes	No direct example in the British Isles. (Crater Lake, Oregon, U.S.A.; 'Maaren', Eifel, West Germany)
5. MAN-MADE	Artificial reservoirs	They make use of natural features especially 'finger lakes'	Lake Vyrnwy, North Wales; Elan Valley, Central Wales

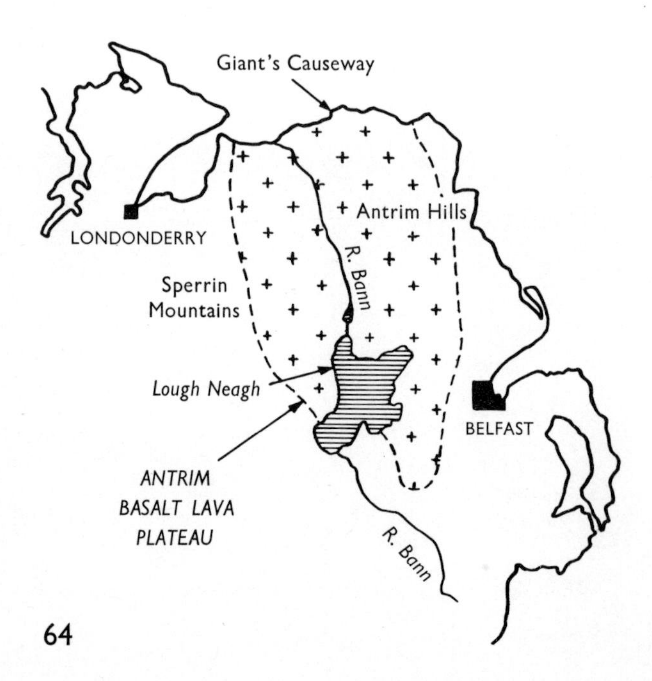

Lough Neagh in Northern Ireland is the largest lake in the British Isles, having an area of 153 square miles. It is a *depression lake*, for it occupies a 'hollow' caused by the collapse of a corner of the basalt plateau as a result of earth movements, and covers much of County Antrim (Fig. 41). The River Bann flows through it and is its sole outlet northward to the sea.

A *lagoon* (Plate XXIX) is a coastal lake. It is separated from the open sea by the sand or shingle bank of a spit or bar stretching across a bay or the mouth of a river. Examples are found in the Norfolk Broads and at Chesil Bank near Portland Bill, Dorset.

Fig. 41 A simple sketch-map to show the location of Lough Neagh

40 Lake Deltas and Flats

The sudden check to a river's flow when it enters a lake causes deposition of part of its load, and a small *lake delta* is formed. A delta formed by a stream entering the side of a lake may extend across the lake to divide it in two. Two such deltas forming on opposite sides of a lake may merge to form an *alluvial flat*. A notable example is seen on the Tourist Map of the Lake District between Lake Bassenthwaite and Derwentwater. These two lakes were formerly one, but they are now divided by a three-mile alluvial flat (Plate XXX).

Natural lakes are but a temporary feature along the course of a river, checking its speed and leading to the deposition of sediment which eventually fills in the lake.

PLATE XXIX A lagoon lake at Loe Bar, Helston, Cornwall

41 Uses of Lakes

Lakes are of great value to man. In the British Isles they form a ready source of water-supply and in North Wales and the Highlands of Scotland they are used as storages for the generation of hydro-electric power. Lakes act as 'filters' for rivers and they can regulate a river's flow, so preventing flooding downstream. Throughout the British Isles lakes which are accessible and possess attractive settings are being utilized as centres of relaxation, and add to the country's tourist attractions.

PLATE XXX A lake flat and delta, between Lake Bassenthwaite and Derwentwater

KEY

1 Lake Bassenthwaite
2 Derwentwater
3 Lake flats, dividing the two lakes
4 River Derwent entering lake by a delta
5 Site of Keswick hidden by a spur of Skiddaw

PLATE XXXI A finger lake—Thirlmere

KEY
1 Wooded moraine
2 Steep wooded slopes
3 Lake beach
4 Helvellyn screes
5 Lower slopes of Helvellyn

EXERCISES

Map IX, Thirlmere (extract from O.S. 1-inch Tourist Map—Lake District).

The 1/63,360 Tourist Map of the Lake District is one of a number specially made for certain tourist areas of the British Isles. These maps are made in extra large sheets of varying sizes, and differ from the 7th Edition Ordnance Survey Maps in the use of layer-colours and hill-shading to give a three-dimensional relief effect. The boundary of the Lake District National Park area is shown by a thick green line. Contours are shown at 50-foot intervals.

1. By selecting examples from the Ordnance Survey map, complete the following:

Type of lake	*Mode of Formation*	*Named example*
(i) Finger		Lake Thirlmere
(ii)		Lake Derwentwater
(iii)		Red Tarn (east of Helvellyn)

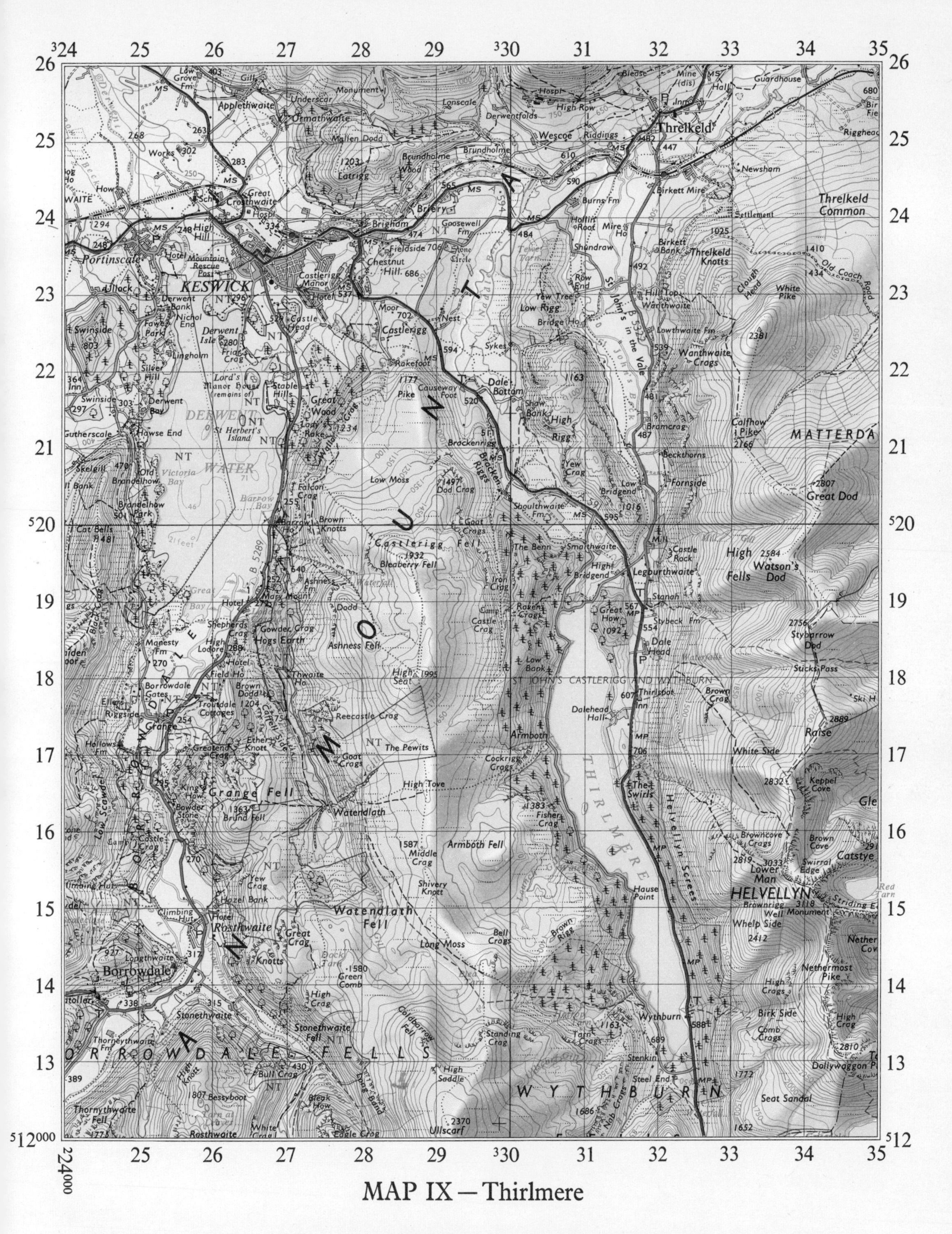

MAP IX — Thirlmere

inted by Lowe & Brydone (Printers) Ltd., London

2. Name the rivers which have contributed to the infilling of the lake flats between Derwentwater and Bassenthwaite (north-west of Derwentwater).
3. Name, and state the height of, the hill at the lower right-hand side of Thirlmere, looking downstream, i.e. to the north. (Refer to both map and photograph.)
4. What kind of valley is that of the 'Helvellyn Gill'? Find its gradient from the 2,650-foot contour line to the point of entry into Thirlmere.
5. Name the relief features in squares 3415, 3319, 2818, 2422, 2925, and 2515.
6. Why are the sides of Thirlmere wooded? State the chief type of woodland in this area.
7. Using your atlas, draw labelled sketch-maps to show:
   (i) the lagoon at Chesil Bank near Portland Bill;
   (ii) the course of the River Shannon, showing Lough Derg and the positions of two other lakes.

   State the cause of each of the above types of lake.
8. Number the road on the east side of Thirlmere and the road on the east side of Derwentwater, and compare their positions.
9. Show how the construction of the railway which crosses the northern half of the map from west to east is related to the relief.
10. Write a brief description of the site of Keswick. Why is there little evidence of other settlement on the map extract?
11. Quote evidence of (i) land use, and (ii) occupations, on the map sheet.
12. Draw an annotated section from High Seat Triangulation Pillar (1,995 feet) at 287180 to Sticks Pass at 341181. Measure the distance in a straight line in miles and yards, and decide whether these two points are intervisible.

APPENDIX

Summary of Landscape Features to be Observed on Ordnance Survey Maps

Having now studied some of the different landscapes of the British Isles and the more important aspects of the physical geography of such regions, you should be familiar with the features which are common to other similar landscapes, and which can be identified on Ordnance Survey maps. Such features are summarized in the Table on the next two pages.

Chalk Escarpments	*Jurassic Limestone Escarpments*	*Clay Vales*
(*a*) Marked scarp and dip slopes. If the scarp faces north-west, then it could be the Chilterns; if south, then it is the North Downs; if north it may be either the South Downs or the Yorkshire Wolds	(*a*) Typical of the Cotswolds, Northampton Heights, Lincoln Edge, and the North York Moors	(*a*) In Lowland Britain clay vales alternate with Chalk and Jurassic (oolitic) Limestone escarpments
(*b*) Smooth, rounded nature of relief forms	(*b*) Features are similar to Chalk, but on a larger scale	(*b*) Generally the relief is low and subdued with gentle slopes
(*c*) Crest of scarp summit has an elevation of from about 650 feet to 850 feet	(*c*) Main differences are: (i) higher elevations of Jurassic Limestone—occasionally over 1,000 feet (ii) the more prominent scarp slope in the case of the Cotswolds, where several streams may flow. This surface drainage is explained by the higher elevations, the higher rainfall (due to the more westerly position), and higher water table (iii) the steep-sided valleys formed by scarp-face streams cutting back into the jointed Jurassic Limestone	(*c*) There are numerous surface streams, for clay is impermeable
(*d*) Prominent river and wind gaps	(*d*) Industrial activities—stone quarries and iron mines, the latter especially in Northamptonshire and Lincolnshire	(*d*) Often low-lying clay areas are liable to flood, hence flood-control drainage schemes
(*e*) Complete absence of surface drainage		(*c*) Industrial activities—clay pits and brick-works
(*f*) Numerous dry valleys		
(*g*) Spring line—sites of farms and settlements		
(*h*) Industrial activities—chalk pits, cement works, lime works, pumping stations		
(*i*) Evidence of past occupation: like many other upland areas in the British Isles, the chalk escarpments were occupied by prehistoric man, as indicated by such terms as Tumuli, Barrows, Entrenchments, and Celtic Fields *Note:* use place-names with care—some can be misleading		

Glaciated Uplands	*Mountain or Carboniferous Limestone*	*Sandstone*
(*a*) These are recognized on O.S. maps by some or all of the following: (i) U-shaped valley—note that in many of these valleys the steep walls may be receding, but in all such valleys the flat floor and misfit stream should be prominent (ii) Beheaded spurs are suggested by the straightened pattern of the contour lines (iii) Hanging valleys—occasionally with waterfall (iv) Cirque with tarn (v) Arête or precipitous rock outcrops (vi) Pyramidal peaks (vii) Ribbon-shaped lakes (*b*) Industrial activities—hydro-electric power installations, electro-metal works, and reservoirs	(*a*) Relief is bolder, more rugged, and higher than in Chalk and Jurassic Limestone landscapes (*b*) Elevations may exceed 2,000 feet (*c*) There is no continuous surface drainage, but streams disappear via pot-holes or swallow-holes (*d*) The term 'scar' may indicate a limestone outcrop, while place-names with 'cove' or 'cave' are usually associated with stream emergence from the limestone (*e*) Industrial activities—lead mines and quarries	*Sandstone Uplands* (*a*) Resistant sandstone uplands, such as the Millstone Grits of the Pennines and Pennant Sandstone of the South Wales coalfield, are areas of bold relief (*b*) Upland summits are level—often plateau-like—and elevations may exceed 1,500 feet (*c*) There are numerous surface streams *Sandstones in Lowland Britain* These occur mainly in Norfolk and the Weald (*a*) There is little surface drainage owing to the porous nature of these less resistant sandstones (*b*) Yet note that the Lower Greensand Ridge (immediately south of the North Downs and the Vale of Holmesdale) forms a prominent ridge culminating in Leith Hill (965 feet) (*c*) On the low sands there may be much heathland with coniferous thickets, or parkland (*d*) Sandstone areas are utilized as residential sites, military training areas, airports, and golf courses (*e*) Industrial activities—gravel and sand pits (*Note* that place-names are here more reliable pointers to rock type than in any other form of landscape)

INDEX